Planetary Science

Developed at
The Lawrence Hall of Science,
University of California, Berkeley
Published and distributed by
Delta Education,
a member of the School Specialty Family

© 2012 by The Regents of the University of California. All rights reserved. No part of this book may be reproduced or transmitted in any form or by any means, electronic or mechanical, including photocopying or recording, or by any information storage and retrieval system, without permission in writing from the publisher.

1372738
978-1-60902-618-9
Printing 2 — 7/2013
Quad/Graphics, Versailles, KY

Table of Contents

Readings

Investigation 2: A Round, Spinning Earth
The First Voyage of Columbus............3
Eratosthenes: First to Measure Earth.......7

Investigation 3: Seasons
Seasons on Earth.....................10

Investigation 4: Moon Study
Lunar Myths......................14

Investigation 5: Phases of the Moon
Measuring Time with Calendars........19
Calculating the Observance of Ramadan..22
Earth's Moon.....................24

Investigation 6: Craters
Craters: Real and Simulated...........31
The Impact That Ended the Reign
    of the Dinosaurs..................36
Gene Shoemaker: Astrogeologist........42

Investigation 7: Beyond the Moon
The Cosmos in a Nutshell..............45
How Earth Got and Held onto Its Moon...54

Investigation 8: The Solar System
A Tour of the Solar System..............58

Investigation 9: Space Exploration
Hunt for Water Using Spectra...........68

Investigation 10: Orbits and New Worlds
Finding Planets outside the Solar
    System........................71

Images and Data............77

References
Science Safety Rules.................133
Glossary..........................134
Index............................136

The First Voyage of Columbus

During the time of Christopher Columbus (1451–1506), European traders made long and difficult journeys across thousands of miles of land. They traveled to the Far East including China, India, Persia, Japan, and Southeast Asia. Europeans called this whole region the Indies, which means the land of many riches. Some of the goods they searched for were spices, gold, and silk. The cost of bringing spices, gold, and silk from the Indies to Europe was high because of the long and difficult journey.

Columbus was the business agent for several important Italian families. He thought that if he could find a sea route to the Indies, he could reduce the time and expense of transportation across land. He would be able to sell his goods at a much higher profit. He could become the richest merchant of all!

In 1488, the Portuguese captain Bartolomeu Dias (1450–1500) sailed down

Christopher Columbus

the west coast of Africa, around the southern tip (the Cape of Good Hope), and up its eastern coast to the Indies. This was known as the eastern route to the Indies.

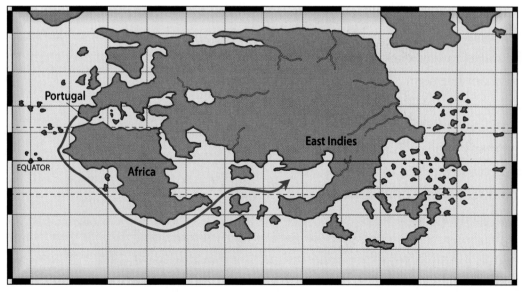

The eastern route to the Indies

Investigation 2: A Round, Spinning Earth **3**

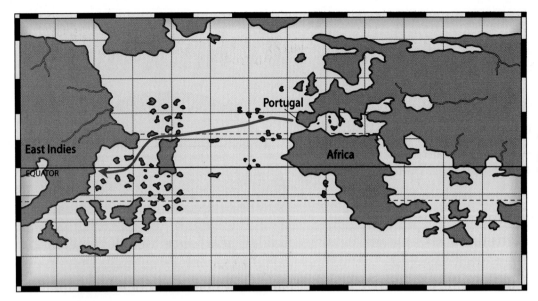

Columbus's proposed route to the Indies

But Columbus had another plan. He wanted to sail west across the ocean to reach the Indies from the other side.

In Columbus's time, the world known to the Europeans did not include the Americas. They had no idea that North America, South America, Central America, or the Caribbean Islands existed. Europeans thought that the great Ocean Sea to the west of Portugal extended west an undetermined distance.

Many people believe that Columbus wanted to prove that Earth is round. But Columbus already knew Earth is round. He had studied the different forms of scientific evidence. He had seen islands appear to rise from the ocean as he sailed toward them. On a flat Earth, this would not happen.

Columbus tried to convince the kings and queens of England and France to finance his journey. He failed many times. His plan was considered risky because it called for sailing west into unknown waters rather than following the long and safer eastern route around Africa.

When Columbus asked for funds from King Ferdinand (1452–1516) and Queen Isabella (1451–1504) of Spain, they listened. They sent him to Spain's best university to talk with professors to determine if his plan was practical.

In those days, all educated Europeans knew that the world is round. Columbus didn't need to prove it. Scholars had known this for over 1,800 years. So what did Columbus argue about with the professors? It was not about the shape of Earth. It was about the size of Earth.

Columbus thought Earth was only about 30,000 kilometers (km) around. The trip from Europe to the Indies using the eastern route was 19,000 km. Columbus reasoned that the distance across the ocean from Spain sailing west to the Indies was probably fewer than 9,500 km.

The world according to Columbus

4

The professors, on the other hand, argued that Earth was about 38,000 km around. They reasoned that the voyage west across the ocean to the Indies was 19,000 km or more. That distance would take about 3 months to cross.

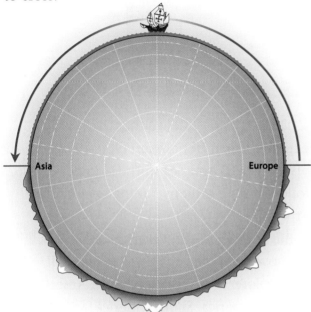

The world according to the professors

That was the problem. The ships of that time could store only enough food and water for a trip of about 1 month. The professors argued that Columbus and his crew would die of thirst or starvation.

Columbus and his crew would be taking a big risk sailing west across the ocean. King Ferdinand and Queen Isabella were also taking a big risk by investing money from the Spanish treasury.

Finally, it was agreed. In 1492, Columbus and his crew set sail in three tiny boats—the *Niña*, the *Pinta*, and the *Santa Maria*—from southern Spain. They sailed south to the Canary Islands, at about 28° **latitude** north, and then straight across the ocean until they arrived at some islands about 2 months later.

Columbus believed that the islands he explored were part of the East Indies. But, as it was determined after the voyages of Amerigo Vespucci (1451–1512), Columbus had landed not in the East Indies, but in the "New World." The islands, in fact, were part of the Bahamas, just off the southern tip of Florida. Columbus was lucky that the islands were there. If he had not struck land when he did, he and his crew would have died of thirst, as the scholars had predicted.

A replica of the *Santa Maria*

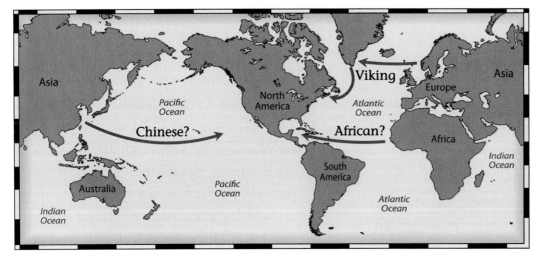

Possible routes of the African, Chinese, and Viking explorers

So did Columbus discover America? Columbus did land in the Americas, but by accident. He was headed for Asia but landed on a continent that he was unaware of. There is speculation that other explorers arrived in the Americas much earlier. Around 2,800 years ago, African sea-faring traders might have sailed to present-day southeastern Mexico. Over 1,500 years ago, the Chinese Buddhist priest Hui Shan is reported to have sailed to the Americas across the Pacific Ocean. He called the land Fusang. Polynesian explorers may have reached South America around 600 years ago.

There is also evidence that the Viking explorer Bjarni Herjolfsson (960–1022) visited North America in the summer of 986, followed soon after by Viking explorer Leif Eriksson (970–1020).

But the fact is, the Americas were not "discovered" by any of these people. All the explorers who reached North America, Central America, South America, and the Caribbean Islands found native peoples who had been living there for thousands of years.

Illustrations in this article are adapted from Who "Discovered" America? *from the Planetarium Activities for Student Success (PASS) Project, volume 10, Lawrence Hall of Science, University of California, Berkeley: 1992.*

Modern view of the world

Think Questions

1. Columbus sailed to the Canary Islands, at 28° latitude north, and then headed west. Eventually, he landed in the Americas. How do you think Columbus was able to cross the ocean without wandering north or south by accident?

2. When Columbus arrived in the Caribbean, people were living on the islands he "discovered." So, who do you think discovered America?

Eratosthenes: First to Measure Earth

Eratosthenes

More than 2,000 years ago, the Greek librarian and mathematician Eratosthenes (pronounced air•uh•TOSS•then•eze) (c276–196 BCE) heard a story that interested him. It was said that in the city of Syene, Egypt, on the Nile River, at noon on June 21, the **Sun** shone directly down an abandoned well and illuminated the dry bottom. This simple observation started Eratosthenes thinking. For that to be true, he reasoned, the Sun would have to be directly over the well. If the Sun were directly overhead, a pole standing perfectly straight up and down right next to the well wouldn't have any shadow at all.

Eratosthenes had reason to believe that the Sun was very far from Earth, so any beam of light striking Earth would be virtually **parallel** to every other beam of light reaching Earth. Therefore, if Earth were flat, the Sun would be directly overhead everywhere at the time it was directly over the dry well. Furthermore, poles placed absolutely vertically in the ground anywhere on Earth would cast no shadow at that time.

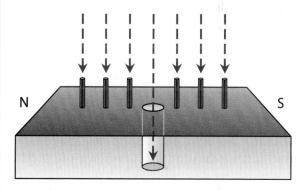

On the other hand, Eratosthenes reasoned, if Earth were round, poles placed straight up and down in the ground would be shadowless in a very small area. Poles would have shadows everywhere else.

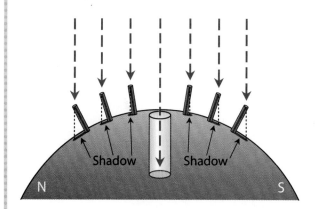

Investigation 2: A Round, Spinning Earth **7**

Eratosthenes discovered that poles outside the city of Syene did cast shadows, and the farther north they were, the longer the shadows. For Eratosthenes, this provided evidence of a round Earth.

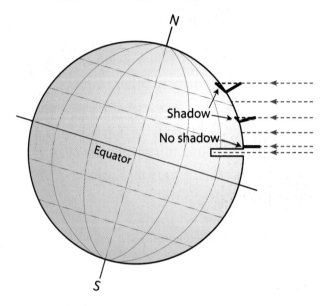

But he was not done. Suppose that Eratosthenes and a friend arranged to observe the shadows cast by identical poles, placed in different locations, at exactly the same time (noon) on June 21. In Syene, next to the well, there was no shadow. The Sun's rays were perfectly parallel to the pole. But in Alexandria, 800 kilometers (km) to the north, there was a shadow. Eratosthenes reasoned that the pole in Alexandria had to be at a different angle than the incoming rays of light from the Sun. Because both poles were perfectly straight up and down, the surface of the land must be at a different angle, like on a curved surface.

After measuring the shadow in Alexandria, Eratosthenes used geometry and a protractor to determine that the Sun shone on the city of Alexandria at an angle of 7.2°.

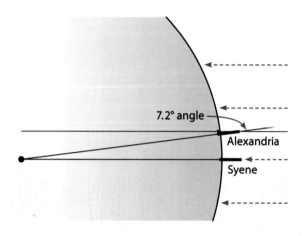

Eratosthenes also knew that distance around an arc (circle) can be described in degrees. A complete circle is 360°. Geometry tells us that 180° describes 1/2 of a circle, and that 90° describes 1/4 of a circle.

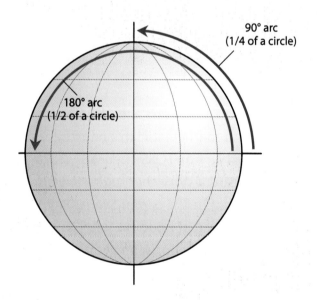

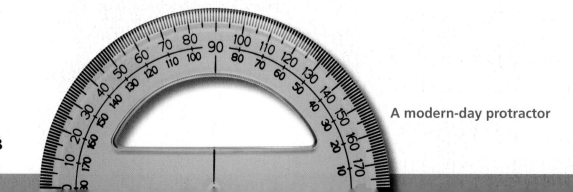

A modern-day protractor

Eratosthenes wanted to know what part of a circle 7.2° represented. Using the same reasoning, he divided 360° by 7.2° and got 50. Now he knew that 7.2° is 1/50 of a circle.

Eratosthenes then measured the distance from Syene to Alexandria in a standard unit used at the time, called the stadium. He found that it was 5,000 stadia between the two cities. So 5,000 stadia was 1/50 of the distance around the circle that represented the **circumference** of Earth. He then multiplied 5,000 by 50, because he knew that the distance between the two cities was 1/50 of the distance around Earth. He found that Earth is 250,000 stadia in circumference.

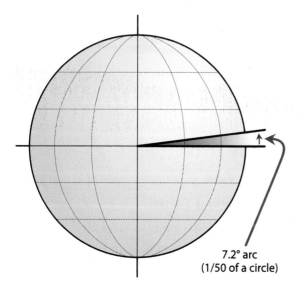

7.2° arc
(1/50 of a circle)

This represents simple and elegant science and math and monumental thinking. Because no one today knows for sure how long a stadium was, it is difficult to argue with the result Eratosthenes calculated. Using a likely value for the stadium of 185 meters (m), modern scholars come up with an estimate of 46,250 km for what Eratosthenes determined the circumference of Earth to be. That's pretty good when compared to today's accepted circumference of 40,074 km.

Today there is no question about the shape of Earth. We've all seen images taken from space, and Earth is round. But think for a moment about the accomplishment of Eratosthenes. More than 2,000 years ago, he measured Earth with some shadows and a protractor.

Think Questions

1. Why do you think Eratosthenes selected June 21 as the date for his observations?
2. What might a shadow look like from an identical pole placed 400 km north of Syene at noon on June 21?

Investigation 2: A Round, Spinning Earth

Seasons on Earth

What do you imagine when you read these words: summer, spring, fall, winter?

Most of us come up with a mental picture or two. Summer means shorts and T-shirts, swimming, and fresh fruits and vegetables. Winter means heavy coats and short days, perhaps with a blanket of snow on everything. **Seasons** are pretty easy to tell apart in most parts of the United States. The amount of daylight, the average temperature, and the behaviors of plants and animals are a few familiar indicators of the season. But what causes the predictable change of season? What have you learned in class that helps you explain the reasons for the seasons?

As Earth Tilts

Let's start with a quick review of some basic information about Earth.

Earth spins on an imaginary axle called an **axis**. The axis passes through the North and South Poles. This spinning on an axis is called **rotation**. It takes 24 hours for Earth to make one complete rotation.

Earth travels around (**orbits**) the Sun. Traveling around something is called **revolution**. Earth's path around the Sun is not exactly round. It is slightly oval. One revolution takes 365 and 1/4 days, which we call 1 year.

 Make a notebook entry. Record the reasons for seasons on Earth. You can add more after reading this article, but record your first ideas now.

North Star

Earth isn't straight up and down on its axis as it revolves around the Sun. It is tilted at a 23.5° angle.

The average distance between the Sun and Earth is about 150 million kilometers (km). Earth's orbit is slightly oval, so Earth is sometimes farther away from and sometimes closer to the Sun. This distance is so insignificant that it is not related to the seasons.

It would seem logical that summer would be when Earth is closest to the Sun. That idea is wrong. Each year when Earth is closest to the Sun, the Northern Hemisphere experiences winter. The reasons for the seasons are linked to Earth's tilt, not the distance from the Sun.

Think about Earth revolving around the Sun. As Earth revolves, it also rotates on its axis, one rotation every 24 hours. Here's something important: Earth's North Pole points toward a reference **star** called the **North Star**. No matter where Earth is in its orbit around the Sun, the North Pole points toward the North Star, day and night, every day all year.

Is Earth closer to the Sun in winter or in summer? Is distance from the Sun a reason for seasons on Earth?

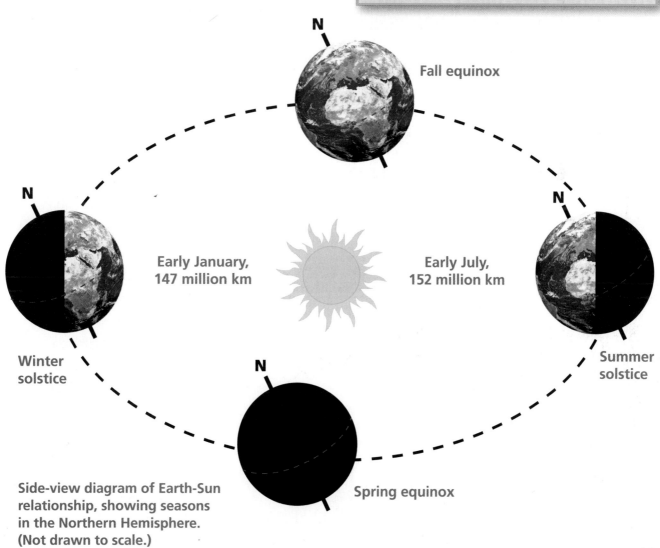

Side-view diagram of Earth-Sun relationship, showing seasons in the Northern Hemisphere. (Not drawn to scale.)

Investigation 3: Seasons **11**

Tilt Equals Season

Look at the illustration on page 11. It shows where Earth is in its orbit around the Sun at each season. You will also see that the North Pole points toward the North Star in all four seasons.

Study the Earth diagram in the summer **solstice** position. Because of the tilt, the North Pole is "leaning" toward the Sun. When the North Pole is leaning toward the Sun, daylight is longer, and the angle at which light hits that part of Earth is more direct. Both of these factors result in more **solar energy** falling on the Northern Hemisphere. It is summer even though Earth is actually farther from the Sun. (And when it is summer in the Northern Hemisphere, it is winter in the Southern Hemisphere.)

Look at the position of Earth 6 months later (at winter solstice). Now the opposite is true. Even though Earth is closer to the Sun at this time, the Northern Hemisphere is tilted away from the Sun. Daylight hours are shorter, and sunlight does not come as directly to the Northern Hemisphere, so it gets less solar energy. It is winter in the Northern Hemisphere.

Four days in the year have names based on Earth's location around the Sun. In the Northern Hemisphere, summer solstice is June 21 or 22, when the North Pole tilts toward the Sun. Winter solstice is December 21 or 22, when the North Pole tilts away from the Sun.

The 2 days when the Sun's rays shine straight down on the equator are the **equinoxes**. On these 2 days, Earth's axis is tilted neither away from nor toward the Sun. *Equinox* means "equal night." Daylight and darkness are equal (or nearly equal) all over Earth. There are two equinoxes each year, the spring (also called vernal) equinox in March and the fall (also called autumnal) equinox in September.

Day and Night

We take day and night for granted. They always happen. Earth rotates on its axis, and the Sun appears to rise; then the Sun appears to set. This cycle has happened at least as long as humans have been on Earth. It will most likely continue for millions of years.

Because Earth is tilted, the length of day and night for any one place on Earth changes as the year passes. This table shows how hours of daylight change at different latitudes during the year. When it's summer in the Northern Hemisphere, the North Pole tilts toward the Sun. During this time at the North Pole, the Sun never sets. Above the Arctic Circle (66.5° north), daylight can last up to 24 hours of the day in the summer. Darkness can last up to 24 hours of the day during the winter.

Think Questions

Go back to your notebook entry about the reasons for the seasons that you made at the beginning of this article. What do you need to add? What do you need to change?

Length of Daylight in the Northern Hemisphere

Latitude	Summer solstice	Winter solstice	Equinoxes
0° N	12 hr.	12 hr.	12 hr.
10° N	12 hr. 35 min.	11 hr. 25 min.	12 hr.
20° N	13 hr. 12 min.	10 hr. 48 min.	12 hr.
30° N	13 hr. 56 min.	10 hr. 4 min.	12 hr.
40° N	14 hr. 52 min.	9 hr. 8 min.	12 hr.
50° N	16 hr. 18 min.	7 hr. 42 min.	12 hr.
60° N	18 hr. 27 min.	5 hr. 33 min.	12 hr.
70° N	24 hr.	0 hr.	12 hr.
80° N	24 hr.	0 hr.	12 hr.
90° N	24 hr.	0 hr.	12 hr.

Lunar Myths

Rona in the Moon

One bright evening, Rona went to fetch water from the stream for her children. In her hand was a basket containing a calabash (a hollow, dry gourd) to hold the water. While she was on her way, the **Moon** suddenly disappeared behind a cloud. On the narrow path lined with trees and bushes, she stumbled on a root. In her momentary anger, she cursed the Moon, saying, "You cooked-headed Moon, not to come forth and shine!"

These words displeased the Moon, who came down to Earth and took Rona. She caught hold of a ngaio tree that was growing on the bank of the stream. The Moon tore the tree up by the roots and flew away, taking Rona, the tree, and the calabash in the basket far up into the sky. Her friends and children, thinking she was away a long time, went to look for her. Not finding any traces of her, they called, "Rona, O Rona, where are you?" She answered from the sky, "Here I am, mounting aloft with the Moon and stars."

When it is a clear night, especially when the Moon is full, Rona may be seen reclining against the rocks, her calabash at her side, and the ngaio tree close by.

Myth from the Maori people of New Zealand. From Myths and Legends of the Polynesians *by Johannes C. Andersen. (New York: Dover Publications, 1995)*

Father Moon

Yasi had a son. One day, the boy was playing with a jaguar and was accidentally killed. The jaguar was terrified and ran deep into the forest to hide forever. Yasi wanted to find out who the killer was, but none of the animals would tell him. This so enraged Yasi that he gave the howler monkey a long neck, put spines on the porcupine, and burdened the tortoise with a heavy shell.

Yasi was still furious, so he leaped up into the night sky to search for his son's killer. He searched all across the land and the sky.

Each day, Yasi got dirtier and dirtier until he was completely covered with soil and leaves. He stopped by a stream to rest and clean himself. Each day he rested, Yasi washed a little bit of his face. Finally, he was completely rested and clean. He resumed the search for his son's killer.

Yasi is still wandering the skies today. He spends half his time hunting and half his time resting. When he returns from the hunt, his face is completely covered in dirt. He washes off a little of the dirt each day until his face shines brightly once again.

Myth from the Siriono people of eastern Bolivia

Moon and His Sister

Moon was a good-spirited and friendly Native American whose face was even brighter than that of the Sun. Moon had one sister, a small star who was often seen beside him. He had many other star friends.

One day Moon gathered all his friends together for a great potlatch. His house was very small. Soon the guests had taken up all the space.

As soon as Moon's sister arrived, Moon asked her to fetch water for him in several buckets. This was not an easy task, as it was winter. She had to fight with the cold, howling wind as she walked to get the water. The water was frozen by the river. She had to chip through the ice to fill the buckets.

She walked back to the house with her heavy burden, only to find that there was no room inside for her. She called to her brother, "Where can I sit?"

Moon was in a very good mood, and he just grinned at his sister. "There isn't enough space for even a mouse in here. I guess you will have to sit on my shoulder," he laughed.

Moon's little sister was not in the mood for his good humor. She took him at his word and jumped onto his shoulder. There she sits even today, holding on to her water buckets. Moon is not as bright as he once was because the shadow from her buckets dims his face.

Myth from the Native Americans of the northwest coast of North America. They held large parties and feasts called potlatches. The host would provide great amounts of food and give extravagant gifts to the guests.

Tale of the Rabbit

There had been four Suns in previous ages, and all had ended in destruction. The gods came together to create a new and final Sun. To create the Sun, one of the gods would have to jump into the fire, the hearth of the gods. The gods gathered at the hearth at midnight to determine which god would make the sacrifice and become the Sun.

Two gods volunteered. One of them was wealthy and strong, the arrogant Tacciztécatl. The other god was poor and sick, the frail Nanahuatzin. Both agreed to die in order to raise the new Sun.

But when the strong god approached the fire, the flames flared high. He drew back from the edge, frightened. His fellow gods shouted encouragements to him to take the leap, but he could not.

Humble Nanahuatzin approached the fire. Without hesitating, he leaped into the flames. His body sizzled, crackled, and burned. This gave Tacciztécatl the courage to jump, and he did. The other gods waited in the darkness for signs of the dawn and Nanahuatzin's transformation into the fifth Sun.

The gods kept arguing among themselves as to where the Sun would first appear. Each one took up a position looking in a different direction so that they would be sure to see the Sun as it first came up. Finally the Sun appeared brilliant and red in the east, the direction of creation and new life. Following the Sun, Nanahuatzin, was a second very bright ball, Tacciztécatl, reincarnated as the Moon. Both balls remained still. To get them to move across the sky, the rest of the gods had to sacrifice themselves in the fire.

But before they died, one of the gods did something to the Moon. He grabbed a rabbit and hurled it into the Moon's bright face, darkening it tremendously. You can still see the imprint of the rabbit on the face of the Moon today.

This is an Aztec myth of a willing sacrifice on behalf of the world's renewal.

Bahloo, Moon Man

Bahloo, the Moon, was lonely high up in the empty sky, so he decided to visit Earth. When the campfires were burning and the girls were dancing, Bahloo came down close to Earth and lowered his shining face to speak to the girls. They were frightened by this bright, round white thing and ran away.

The next night he returned to find two other girls sitting on the riverbank. "How beautiful the moonlight is," sighed one of the girls. This was encouraging to Bahloo, and he decided to come closer to the girls. He broke into a run, puffing and blowing, his big belly shaking. The girls were surprised by Bahloo. They didn't know whether to laugh or shout for help. They ran away a safe distance and stood there staring back at the Moon.

Bahloo's feelings were hurt. He sat down by the riverbank and cried. The girls felt sorry for Bahloo and returned. They invited him to ride in their canoe to the other side of the river. But Bahloo was so big that, when he stepped into the canoe, it rocked and tipped, and then turned over, dumping Bahloo into the water. The round shining Moon sank down, down into the water. His light became dimmer and dimmer. The girls laughed and ran home.

Bahloo was very embarrassed. He climbed into the sky without anyone noticing. He remained hidden for several days. Gradually, he regained his courage and grew round and bright for all to see. But when he remembered the girls and how they laughed at him, he began to get smaller and soon went out of sight. Every month, Bahloo grows round and bright and full of courage, and every month he remembers his fall into the river and shrinks away to hide.

Aboriginal tribal myth, Australia

Measuring Time with Calendars

Calendars organize our lives. Calendars mark important occasions, help us plan future events, and remind us how many days are left until the weekend. How would life change if there were no calendars? How would you celebrate your birthday if you couldn't tell when it was? What did people do before they had calendars to keep track of important days and events that were coming up?

Natural Cycles

Calendars allow us to keep track of the time that has gone by and the time that is coming up. For thousands of years, long before our modern calendar was developed, people observed nature to mark time. One of the most obvious measurements of time is the day. Measuring sunrise to sunrise allows us to calculate time by counting days, so we can keep track of events in the past (such as 10 days ago) and think about events in the future (such as 3 days from now).

Long ago, people noticed that the length of time between sunrise and sunset is not always the same. By observing carefully, day watchers in the Northern Hemisphere determined that one day each year was shorter than all the others. We call that day the winter solstice, which marks the first day of winter. Then days got longer and longer until the longest day of the year, which we call the summer solstice. Days then got shorter until once again winter solstice returned. The time from one winter solstice until the next is 365 1/4 days, or 1 year.

The Moon

Another way people measured time was based on Moon observations. Over time, shepherds and other sky watchers observed the repeating pattern of Moon **phases**, like you have been doing in class. When you tally the days from one **full Moon** to the next full Moon, the cycle is always 29 or 30 days. The unit of time from one full Moon to the next is one lunar month.

Can lunar cycles help us measure a year? If you keep track of the number of lunar months from one winter solstice until the next, you would count approximately 12 lunar months, that is, 354 days. This is called a lunar year.

Lunar Year and Solar Year

People who developed the early calendars most often used lunar cycles. However, the lunar year is not an exact match with the solar year. A solar year can be measured exactly from one winter solstice to the next winter solstice. But in a lunar year, you would find that the winter solstice lands on a different date every year because the lunar year is about 11 days shorter than the solar year. This mismatch has been a problem for calendar makers.

The lunar cycle was one way people measured time in the past.

This chart shows the months of the modern calendar, which measures a solar year, compared to the lunar year. The lunar year is 11 days shorter, leading to a mismatch when calculating solar events with a lunar calendar.

Solar year	January: 31 days	February: 28 or 29 days	March: 31 days	April: 30 days	May: 31 days	June: 30 days
Lunar year	Lunar Month 1: 29 days	Lunar Month 2: 30 days	Lunar Month 3: 29 days	Lunar Month 4: 30 days	Lunar Month 5: 29 days	Lunar Month 6: 30 days

The Modern Calendar

It is no coincidence that 1 month (30 or 31 days) is about the same length as a lunar month, or "moonth" (29.5 days). The calendar that we use today started off by following the patterns of the Moon, but has been changed many times. An early version of our calendar used in ancient Greece measured the beginning of each month with the sighting of the new **crescent Moon**. Over 2,000 years ago, the Roman Empire decided that the calendar should stop trying to match each month with the Moon. After that, all months except February were assigned either 30 or 31 days, and the year officially became 365 days.

The 365-day calendar we use today is called the Gregorian calendar, and its origins are rooted in the Greek and Roman calendars described above. This calendar more accurately predicts annual events like the solstices and is called a solar calendar. The solar calendar depends on Earth's position in its orbit around the Sun. This is the calendar that is used in most of the world for official business.

However, many people use different calendars for religious observances and traditional celebrations. You may have noticed that Rosh Hashanah (Jewish), Ramadan (Muslim), Diwali (Hindu), Easter (Christian), and Chinese New Year do not occur on the same date every year. All of these holidays occur on different calendars that are still connected to the lunar cycle. There are also calendars that follow solar and lunar (called lunisolar) cycles.

A Sample of Calendars Used around the World			
Type of calendar	Lunar	Solar	Lunisolar
Chinese			X
Gregorian		X	
Hindu			X
Islamic	X		
Jewish			X
Persian		X	

While the modern calendar began by following and connecting the patterns of nature, over time we have relied less and less on observing nature. It is more common for us to look at a calendar on a wall or a computer to know the date. The calendar system that we currently use was set over 400 years ago and will probably not change anytime soon.

Think Questions

1. In what ways does the modern calendar connect with nature?
2. Do you or anyone you know follow more than one calendar system? Which ones?
3. What is a leap year? When is it, and why do we have it?

July: 31 days	August: 31 days	September: 30 days	October: 31 days	November: 30 days	December: 31 days
Lunar Month 7: 29 days	Lunar Month 8: 30 days	Lunar Month 9: 29 days	Lunar Month 10: 30 days	Lunar Month 11: 29 days	Lunar Month 12: 30 days

Calculating the Observance of Ramadan

It's April 2020 and it's Ramadan. That means Hussein Malik and his family in Oakland, California, are preparing to fast. Ramadan is a lunar month in the Islamic calendar when Muslims don't eat or drink from sunrise to sunset. This practice is meant to encourage reflection on one's religious practice. The sighting of the new crescent Moon marks the first day of each month on the Islamic calendar, and the ninth month marks the start of Ramadan.

During Ramadan, Hussein's family can eat their first meal of the day after sunset, around 8:00 p.m. But because the Islamic calendar is lunar, the dates change by about 11 days every year. Hussein's mother remembers years when Ramadan started in the winter. Thanks to winter's shorter daylight hours in the Northern Hemisphere, fasting was a lot easier because she finished her fast when the Sun set around 5:00 p.m. Hussein's family started fasting this year on April 26, after getting a phone call from relatives saying that the crescent Moon was sighted in California.

Hussein's cousins Bilal and Aisha live in Dallas, Texas. In Dallas, many people started fasting on April 24, two days earlier. Bilal and Aisha's community in Dallas follows a system that uses scientific calculations to determine the first day of Ramadan. Because people can predict Moon phases, as we have in class, some Muslims use science to signal the start of Ramadan instead of waiting for an actual sighting of the crescent Moon. Astronomers determined that the crescent Moon was above the horizon for about 30 minutes in parts of North America immediately after sunset. The scientists did not actually see the Moon, but their calculations told them that it was in the evening sky in Dallas on April 24.

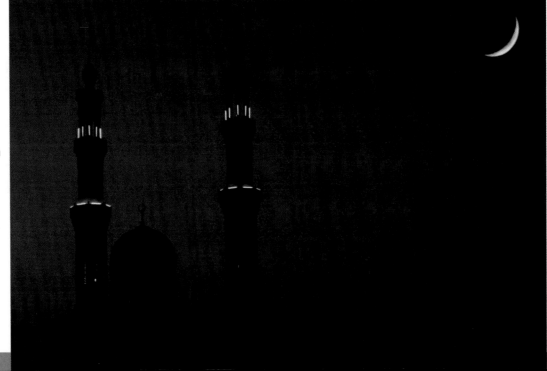

The sighting of the new crescent Moon marks the first day of each month on the Islamic calendar.

How can two families observe the same religious event on different days? Which is the "true" date of Ramadan? Over recent decades, Muslims in North America have been debating how to determine the beginning of the month of Ramadan. One view is to preserve the traditional aspect of Ramadan, which means relying on local observers to see the first crescent Moon so that they can announce the beginning of the month. Weather can affect visibility of the Moon, and different parts of the world have different views of the sky as Earth rotates. Therefore, the month of Ramadan may start on different dates in different places.

But many Muslims consider the direct observation of the crescent Moon unnecessary now. That's because it is possible to know precisely when Ramadan starts based on the rotation of Earth and the location of the Moon in its orbit around Earth. Using scientific calculations also allows Muslims to address modern concerns such as taking time off from school or work for religious observance at the beginning and end of the month. This is one example of science and religious belief working together. Some Muslims use scientific observations of the Moon to inform their religious practice. They don't stop having religious beliefs just because they're using science. In fact, people of many different religious beliefs use and study science every single day.

The dilemma for the Muslim community highlights the different ways in which time can be marked. On the one hand, using the Moon sighting as a way to mark the new month is how people have been marking Ramadan for over 1,400 years. It keeps people connected with the natural cycles of the Moon, something that people all over the world had been doing for thousands of years before the calendar was standardized.

On the other hand, starting the month based on scientific calculations makes the calendar universal. In the modern world, where people rely on numbers to know the time, a standardized calendar makes it easier to live according to an established schedule and allows people to prepare for the month ahead of time. In future years, the Muslim community and other cultures will continue to debate the benefits of traditional versus modern methods of observation.

Think Questions

1. Now that it's easy to share photographs over the Internet, some people use photographic evidence to start Ramadan. Would you trust a photograph posted to a website as evidence to begin the month of Ramadan? Why or why not?

2. Imagine you had to decide for your family which way to start and end Ramadan. Would you follow the Moon-sighting method, or would you use scientific calculations to determine the new crescent Moon? Why did you choose this method?

Investigation 5: Phases of the Moon

Earth's Moon

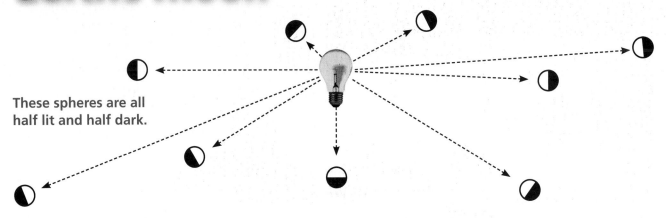

These spheres are all half lit and half dark.

The Moon shines so brightly in the sky that you can sometimes see it even during the day. But the Moon doesn't make its own light. The light you see coming from the Moon is reflected sunlight. The Moon is a sphere. When light shines on a sphere, the sphere is half lit and half dark. It doesn't matter where you position the sphere. It is always half lit and half dark.

The same is true for the Moon. It is always half sunlit and half dark. The half that is lit is the side toward the Sun. The half that is dark is the side away from the Sun. It takes about 4 weeks for the Moon to orbit Earth. Look at the diagram below to think about where you are on Earth when you see the Moon during day or night.

The Moon doesn't always appear to be the same shape. That's because half of the Moon is always dark. The other half is lit by the Sun. As the Moon orbits Earth, observers on Earth see different amounts of the lit half. The different shapes of the Moon are called phases. The phases change in a regular pattern as the Moon orbits Earth. The Moon completes an orbit and goes through its cycle of phases in just over 4 weeks.

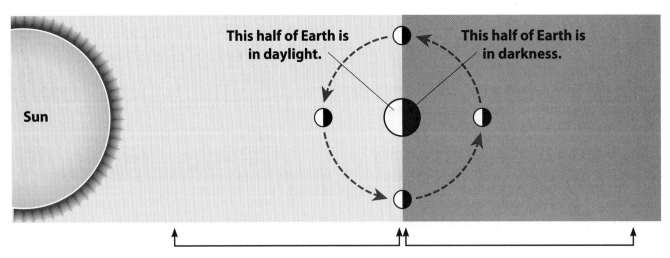

When the Moon is in this half of its orbit, we see the Moon during the day.

When the Moon is in this half of its orbit, we see the Moon during the night.

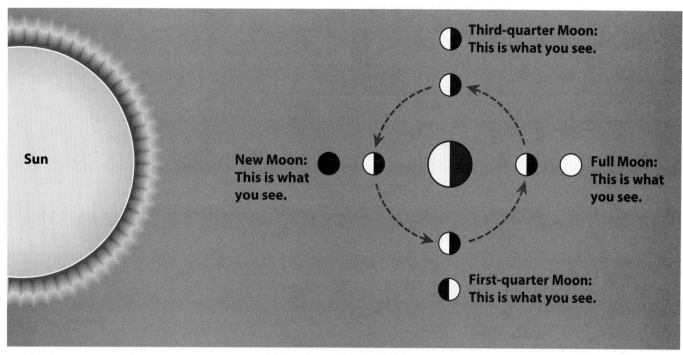

The inner circle shows the positions of the Moon in its orbit.

Four specific phases happen about 1 week apart. The **new Moon** is invisible to us. It occurs when the Moon is between Earth and the Sun, so we are looking at the dark side of the Moon. The full Moon occurs when Earth is between the Sun and the Moon, so we are looking at the lit side of the Moon. Halfway between the new Moon and the full Moon is the **first-quarter Moon**, so we see half lit and half dark. Halfway between the full Moon and the new Moon is the **third-quarter Moon**.

The Moon changes through its phases in a predictable pattern. The first appearance of the Moon after the new Moon is a thin crescent. The next day the crescent will be a little bigger. The Moon has moved in its orbit so that we can see a little bit more of the lit part, and a little bit less of the dark part. The crescent will get bigger each day until the first-quarter Moon. Getting bigger is called **waxing**. After the first quarter, the Moon continues waxing. But it is no longer a crescent Moon. It is a **gibbous** Moon. The gibbous Moon is nearly round on the side facing the Sun, with a bulge on the other side. The gibbous Moon waxes until it appears completely round. That is the full Moon.

Investigation 5: Phases of the Moon **25**

For the next 2 weeks the Moon is **waning**. Each day it appears to be a little smaller. The Moon has moved in its orbit so that we can see a little bit more of the dark part and a little bit less of the light part. The waning gibbous Moon becomes the third-quarter Moon. Then the Moon becomes a waning crescent Moon. At the end of just over 4 weeks, the lunar cycle is complete, and the Moon is new again.

On your Moon log, you drew the appearance of the Moon for a month. If you repeated the process over and over, that is what you would see each cycle. That is the lunar cycle that you will observe month after month.

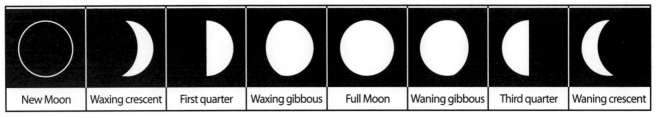

Lunar-month drawings

Lunar-month photos

A total lunar eclipse

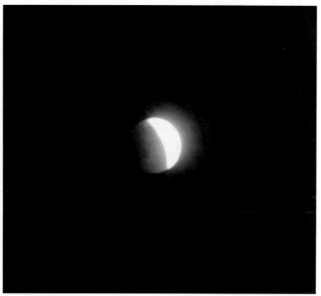
A partial lunar eclipse seen from Earth

Eclipses

Occasionally people on Earth are able to observe a lovely orange-colored eclipse of the Moon (a **lunar eclipse**). Even less frequently they can observe a black-centered eclipse of the Sun (a **solar eclipse**).

What causes these interesting events? When can you see a lunar eclipse? When can you see a solar eclipse?

A total solar eclipse

A solar eclipse seen from Earth

Investigation 5: Phases of the Moon

Solar eclipse. A total solar eclipse occurs when the Moon passes exactly between Earth and the Sun. The Moon completely hides the disk of the Sun when this happens.

This is how Earth, the Moon, and the Sun are aligned when a solar eclipse is observed. The place where a solar eclipse can be observed is restricted to a very small location on Earth's surface. A total eclipse of the Sun can be observed for several minutes as the disk of the Moon passes across the disk of the Sun.

A solar eclipse seen from Earth

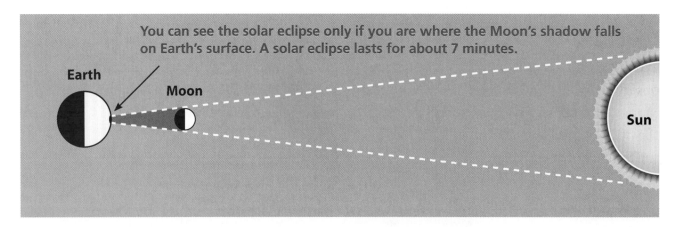

You can see the solar eclipse only if you are where the Moon's shadow falls on Earth's surface. A solar eclipse lasts for about 7 minutes.

The Moon travels around Earth in about one month. Why doesn't a solar eclipse occur every month? The Moon's orbit around Earth is not in the same plane as the orbit of Earth around the Sun. The Moon's orbit is tilted a little bit, so most months the Moon, the Sun, and Earth do not align exactly.

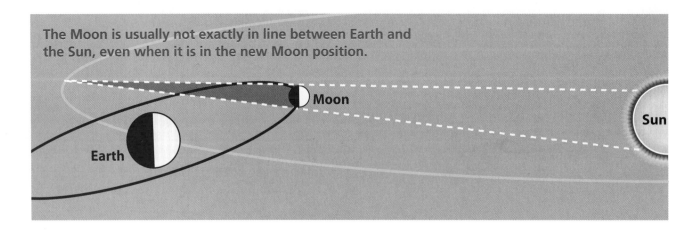

The Moon is usually not exactly in line between Earth and the Sun, even when it is in the new Moon position.

Lunar eclipse. An eclipse of the Moon occurs when Earth passes exactly between the Moon and the Sun. The Moon moves into Earth's shadow during a lunar eclipse. At the time of a full lunar eclipse, Earth's shadow completely covers the disk of the Moon. This is how Earth, the Moon, and the Sun are aligned for a lunar eclipse to be observed.

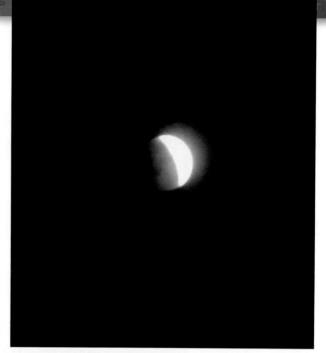

A partial lunar eclipse seen from Earth

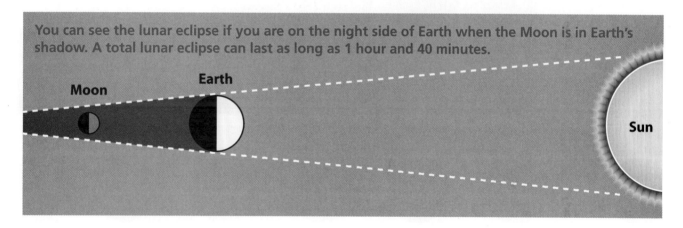

You can see the lunar eclipse if you are on the night side of Earth when the Moon is in Earth's shadow. A total lunar eclipse can last as long as 1 hour and 40 minutes.

Why don't we see a lunar eclipse every month? Because of the tilt of the Moon's orbit around Earth, Earth's shadow does not fall on the Moon in most months.

And to make things even more complicated, the orientation of the Moon's orbital plane changes a little bit each year.

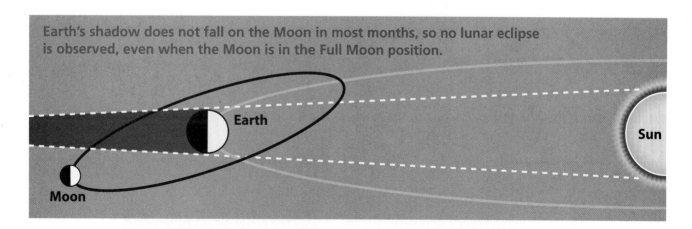

Earth's shadow does not fall on the Moon in most months, so no lunar eclipse is observed, even when the Moon is in the Full Moon position.

Investigation 5: Phases of the Moon

The sequence of images below shows the Moon as it is being eclipsed. The images show the Moon's movement into Earth's shadow. Below the images is a diagram showing how the Moon, moving from right to left in a counterclockwise orbit around Earth, passes into Earth's shadow, and ends up completely eclipsed. The reddish-brown color of the last image is typical of a fully eclipsed Moon.

Why is the fully eclipsed Moon reddish-brown and not invisible? When light passes through Earth's atmosphere, it is bent and scattered by the air particles. As a result, instead of ending up in a completely black shadow, some reddish light falls on the Moon's surface, making it appear reddish-brown. If this bending and scattering did not occur, a totally eclipsed Moon would be invisible because no light would hit the Moon and no light would be reflected back into our eyes on Earth.

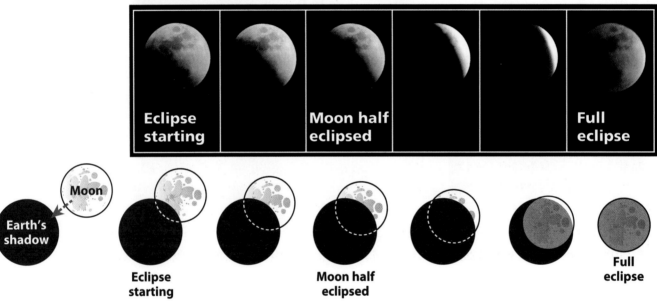

The Moon's movement into Earth's shadow

Think Questions

1. During what phase of the Moon can you observe a lunar eclipse?
2. During what phase of the Moon can you observe a solar eclipse?

A total lunar eclipse visible from Miami, Florida

A total solar eclipse

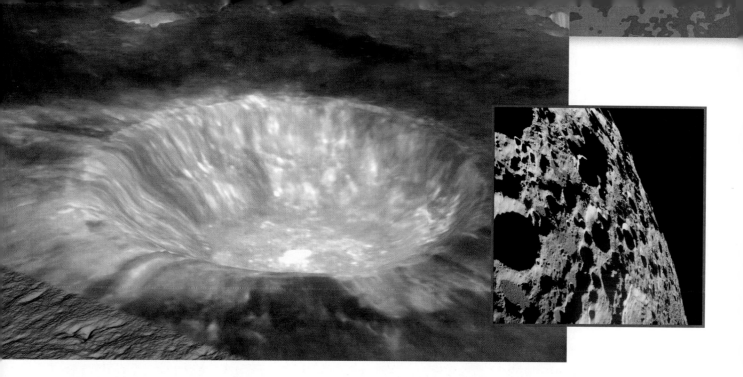

Craters: Real and Simulated

What created the **craters** on the Moon? Was it primarily volcanic activity or impacts? For years, scientists were not sure. The debate in the early 1960s had two well-known advocates: geologists Jack Green (1925–) and Gene Shoemaker (1928–1997). Green thought the craters were inactive volcanoes. Shoemaker thought the craters were scars created by objects that slammed into the Moon's surface.

A piece of important information that helped decide the debate was the discovery on Earth of a type of quartz called shocked quartz found in underground test sites for nuclear bombs. Shocked quartz is formed by tremendous pressure, like that generated by a nuclear explosion. Later, Shoemaker found shocked quartz inside the Barringer Crater (also known as Meteor Crater) in Arizona. The presence of shocked quartz confirmed that this crater was formed by an impact. A volcano would not generate the pressure required to produce shocked quartz. Green and Shoemaker agreed that they could decide for sure which process produced the craters on the Moon if they could examine Moon rocks from the central peaks in some of the larger craters. The successful Apollo missions provided those Moon rock samples that confirmed the impact theory.

When the **solar system** was young, over 4 billion years ago, a huge number of small objects flew around and crashed together. Sometimes the objects broke into smaller pieces. Other times they stuck together. In the first half billion years of the solar system, violent collisions were so frequent that there was no chance of life on any **planet**.

Even today, countless small and medium-sized pieces of rock and metal, called **meteoroids**, orbit the Sun. Every day, tiny particles rain down on Earth, some as small as grains of dust. Most of them slow down and burn up in Earth's **atmosphere**. Slightly larger ones, gravel- and pebble-sized, make streaks of light in the night sky that we call **meteors**, commonly known as shooting stars.

An artist's rendition of an asteroid about to hit Earth

Larger objects move right through the atmosphere as easily as a stone through a spiderweb. Some of these are relatively small, and some are the size of cities. These larger objects are called **asteroids**. Many asteroids are orbiting the Sun between the orbits of Mars and Jupiter in a region called the **asteroid belt**. Many other asteroids have orbits that can take them through the inner solar system. When a meteoroid or an asteroid hits a moon or a planet, it creates a crater. Tiny meteors make craters that are microscopic. Large asteroids can produce huge craters that would take a person many hours or even days to walk across.

Occasionally, icy objects, called **comets**, come flying through the inner solar system. Comets orbit the Sun, but they spend most of their time way out beyond the outermost planets in the solar system in a region called the Oort cloud. The Oort cloud is named after a Dutch astronomer, Jan Hendrik Oort (1900–1992). A comet can take decades or even hundreds of years to complete one orbit. A comet might end up on a collision course with a moon or a planet. Comet Shoemaker-Levy 9 hit the planet Jupiter in 1994 after Jupiter's **gravity** ripped it apart into more than 20 pieces. If a comet or an asteroid of this magnitude were to hit the Moon, the resulting crater would be a major feature on the Moon's surface.

A comet

Scientists in laboratories and students in classrooms can simulate impacts in order to study crater formation. They might use sand or flour to simulate the Moon's surface and marbles or rocks to simulate meteoroids. The "meteoroids" dropped on the surface material create impacts. The resulting craters have the characteristic hole, rim, **ejecta**, and rays of natural craters, but one element is lacking. Even when scientists simulate crater formation by firing objects into the sand at very high speeds, the speed of the objects is far slower than the pieces of debris traveling through space, typically at tens of thousands of kilometers (km) per hour. It's impossible for the energy released in the laboratory or classroom at the moment of impact to accurately represent what really happens when a meteoroid or an asteroid slams into the Moon.

A meteoroid traveling 72,000 km per hour (20 km per second) crashing into the Moon will cause major damage. The impact releases a tremendous amount of energy. The force of the impact creates so much pressure and heat that the meteoroid can vaporize, explode, and disappear. It is the explosion that creates the crater by blasting the soil away in all directions.

One way that scientists have studied crater formation is to observe the results of explosions at bomb test sites in large expanses of sand. These events are similar to what happens when a meteoroid impacts, vaporizes, and explodes.

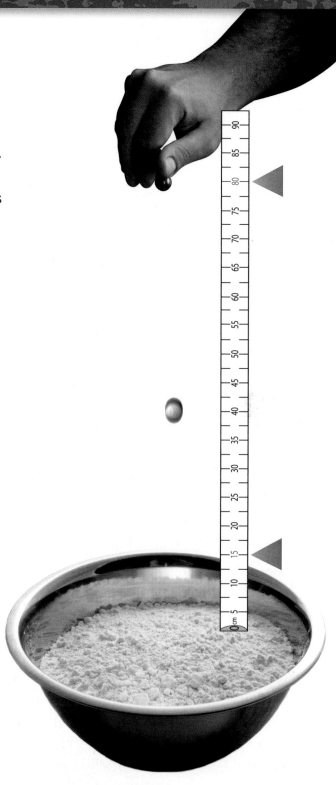

Dropping marbles into flour is one way you can simulate crater formation on the Moon's surface.

In your notebook, review the procedure and findings of the crater experiment from class. Make a new notebook entry explaining how your experiment was similar to and different from an actual meteoroid impact.

Investigation 6: Craters **33**

Simple and Complex Craters

Small explosions produce small bowl-shaped craters with a fairly uniform blanket of ejecta distributed around the rim. These are called **simple craters**.

A large terraced crater

Simple craters

Larger explosions produce so much pressure below the point of the impact that there is a rebound effect in the center of the crater. The shock wave after the explosion often pushes up a big mountain in the middle of the crater. **Complex craters** often have central peaks and ejecta thrown out in long rays.

Really big, complex craters have an additional feature called terraces. They look like giant steps leading from the crater floor up to the rim.

In the past, the Moon had a molten core. Sometimes, an impact was so huge that it cracked the outer layers of the Moon. Magma would seep up into the crater and result in a **flooded crater**. When the magma cooled, the dark surface of the cooled rock became smooth and uniform, resulting in the feature we call **mare** (plural *maria*). Later impacts, however, might have left new marks in the mare.

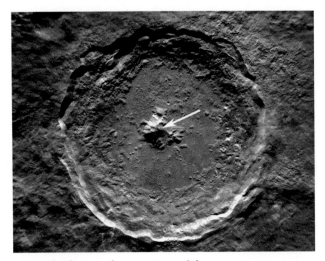

A typical complex crater with a central peak

Can you see the circular shape of the original impact crater? Can you see craters that have since formed on the dark surface of the mare?

Comparing Models to Actual Craters

In class, you simulated the formation of craters on the Moon by dropping marbles into flour. You produced craters, similar in many ways to those seen on the Moon. How are these craters actually like real Moon craters, and how are they different?

Size of Craters. Your craters are 2–3 centimeters (cm) in **diameter**. The smallest Moon craters are microscopic. The largest one, the Aitken basin at the Moon's south pole, is 2,500 km across and 13 km deep. It would cover most of the United States! The Imbrium basin (Mare Imbrium), another large impact crater, is 1,200 km across.

Size of Projectiles. Your marbles are approximately 1.5 cm in diameter. Scientists believe the asteroid that collided with the Moon to create the Imbrium basin was 100 km in diameter. A meteoroid as large as New Hampshire can create a crater almost as large as the continental United States. The objects that made the craters on the Moon came in a range of sizes.

Speed of Projectiles. Dropped from a height of 200 cm, or 2 meters (m), your marbles reach a speed of perhaps 1 km per hour. The asteroids and meteoroids that struck the Moon would have been traveling at speeds of about 72,000 km per hour! Unlike Earth, the Moon has no atmosphere to slow a meteoroid as it approaches the surface.

Impact Event. When your marbles strike the flour, a "splash" of flour sprays out across the surface of the pan. On the Moon, due to the speed and energy of large meteoroids, the impact is an explosive event. The heat generated by the impact is so intense that the rock and the meteoroid instantly vaporize. The resulting explosion caused by expanding gases blasts a huge hole and throws debris in all directions. This is how a 100-km object can create a crater with a diameter of 1,200 km.

Draw a line of learning under your notebook entry comparing the crater experiment to an actual impact, and add ideas using information from this article.

Think Questions

The photo on the left is Barringer Crater in Arizona. Think about how it might have formed.

1. Is it a simple or complex crater?
2. How old do you think it might be? Why?
3. Why do you think there are so few craters on Earth and so many on the Moon?

Investigation 6: Craters

The Impact That Ended the Reign of the Dinosaurs

Scientists have had many debates about what ended the reign of dinosaurs. Dinosaurs are usually studied by paleontologists, through fossil remains. Normally, changes in populations of plants and animals take place very slowly. But in certain periods of history a large number of species died at about the same time in events called mass extinctions. The last time a mass extinction occurred was 65 million years ago, when the dinosaurs became extinct, along with 50 percent to 75 percent of all other species. The Age of Reptiles ended, and the Age of Mammals began.

The debate is about how fast this event occurred and what caused it. It was likely due to climate change, but there have been two competing theories about the cause of the climate change. There is evidence for both theories: active volcanism and a huge asteroid or comet impact, both about 65 million years BCE, separating the Cretaceous and Tertiary periods. Either one or both of these events would have had a profound effect. Most species did not survive the event. It took many thousands of years for a diversity of life-forms to repopulate Earth.

Dinosaurs ruled Earth for 165 million years.

Dinosaurs ruled Earth for 165 million years. Their fossils are found in sedimentary rocks that formed when layers of sand, dust, clay, and ash piled up. Over time they turned into rock. Deep layers of rock are older than the layers above. Older fossils are found in the deeper layers of rock. The fossil record shows that about 65 million years ago, all traces of dinosaurs vanished.

Looking for Evidence

In the 1970s, a scientist named Dr. Walter Alvarez (1940–) was investigating sedimentary rocks on a mountain in Italy. He found an unusually high concentration of an element called iridium. While iridium is rare on Earth's surface, he knew that some meteoroids and asteroids are rich in iridium. Alvarez began to wonder if the layer of iridium he discovered in the rocks could be asteroid or **meteorite** dust.

In the laboratory, analysis showed that the thin layer of rock in which the iridium occurred was 65 million years old. Alvarez reasoned that there might have been a big asteroid impact on Earth 65 million years ago that accounted for the iridium in his rock samples.

Alvarez and his research group wondered if an iridium layer would be found in 65-million-year-old sedimentary rocks from other places on Earth. Dust and debris from a large impact would likely be scattered far and wide. When they looked, they did find high concentrations of iridium in rocks of the same age in many other places on Earth.

Then they wondered what process might have laid down iridium all over Earth at the same time. The answer that they came up with was one huge impact that blasted a huge amount of dust, including iridium, into the atmosphere. The dust would have circled the globe for months, carried by circulating winds, then settled out of the atmosphere all over Earth.

Dr. Walter Alvarez and his wife

It wasn't long before Alvarez and his group wondered if the extinction of the dinosaurs and the suspected asteroid impact could be related, since both happened about 65 million years ago. But how could one asteroid impact cause the extinction of dinosaurs all over Earth?

An impact on Earth would have dramatic effects. Assume the asteroid was 10 km wide. The asteroid or comet would vaporize as it blasted a crater 160 km wide and 8 km deep. It would release energy equivalent to a billion-megaton nuclear bomb (or as Alvarez estimated, equivalent to the energy in all nuclear weapons in all arsenals on Earth). The blast would send a huge amount of material into the atmosphere and into space. Shock waves from the blast would create tsunami waves several thousand meters high. The waves would travel hundreds of kilometers and drown and crush everything in their paths. The blast would also create winds of hundreds of kilometers per hour, and storms would have raged for weeks. Intense heat from the blast and from hot ash raining back on Earth would start forest fires that could burn as much as 50 percent of Earth's forests. The intense heat and pressure would also cause chemical reactions and create acid rain. Many of the remaining organisms in acidic waters would be killed.

Perhaps most devastating of all would be the huge cloud of very fine material ejected from the impact site. The dust and smoke from worldwide forest fires would block out sunlight. Alvarez and other scientists worked up some scientific models that suggested that an asteroid about 10 km across could throw enough material into the atmosphere to block the Sun for a year. Temperatures would have dropped to freezing. Most plants and animals would have died. Animals that survived the

impact would have died when they found nothing to eat. Only a few of the hardiest organisms that were lucky enough to be in a few protected areas would have survived.

There was one problem with this theory, however. The event should have left a very large crater. A 10-km asteroid would produce a crater at least 160 km across. Years of searching uncovered no evidence of such a crater. Many suggested that the impact site might be hidden under the ocean, which covers about two-thirds of Earth's surface.

Since geologists first started a search for the impact crater that killed the dinosaurs, new technologies have emerged. Scientists now have advanced magnetic sensors as well as satellite sensors that can "see" through water to look for unique, ancient features of the land. An interesting structure was discovered in the Gulf of Mexico.

The colorful image on the right may not look like much to the untrained eye, but now that you have studied craters and know what to look for, you can see the distinctive shapes that suggest a crater.

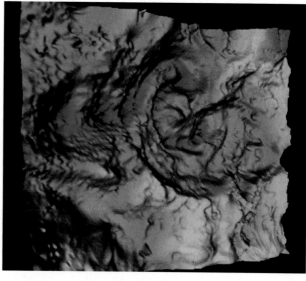

A computer-generated map showing the shape of the basin of Chicxulub Crater

The picture below is a shaded relief image of the northwest corner of Mexico's Yucatán peninsula, provided by the Shuttle Radar Topography Mission (SRTM). Color indicates elevation of the land. The green is at the lower elevations, rising through yellow and tan to gray at the highest elevations. The dotted line shows where the crater rim is below the surface rocks that formed long after the impact.

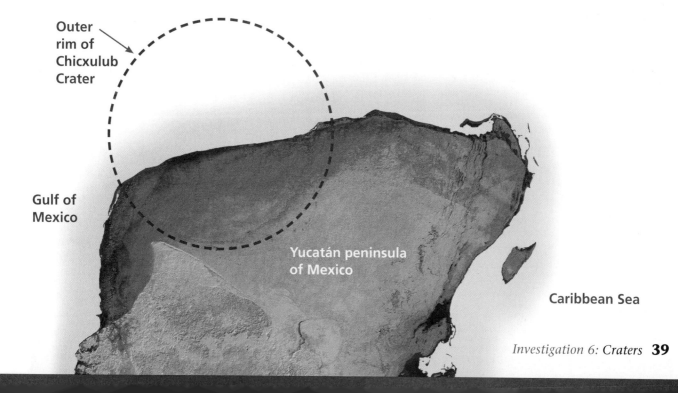

Investigation 6: *Craters* 39

This structure is named Chicxulub (pronounced cheek•shoe•lube) Crater, because its center is located near the town of Chicxulub, on the Yucatán peninsula of Mexico. The crater is one of the largest impact structures in the world. It is more than 180 km in diameter and more than large enough for the impact predicted by the Alvarez team.

Glen Penfield, a geophysicist working for the Mexican oil company Pemex, discovered the crater in 1978. Rock samples from the crater site contained shocked quartz, good evidence for the impact origin of the crater. Further investigation of the crater site revealed a gravitational distortion and tektites (glassy pieces usually found in meteor impacts).

What do you think? Could the dust and smoke thrown into the atmosphere from this impact have darkened the skies for months and killed many plants and animals?

Although the scientific community often engages in debates in search of truth, they also give recognition and awards for work well done. In November 2008, Walter Alvarez received the prestigious Vetlesen Prize for his conclusions about asteroid impacts. The Vetlesen Prize is an award for earth sciences that is comparable to the Nobel Prize. Alvarez's work changed the way scientists view the history of Earth and challenged the long-standing belief that everything that has happened in the history of Earth happened slowly and gradually.

Future Earth Impacts

Earth and the Moon share a similar history in the solar system. Both have been in the paths of asteroids and comets since they formed about 4.6 billion years ago. Craters on Earth don't last very long in geological time because they are erased by erosion from wind

and moving water. Craters on the Moon have not been erased by wind or water because there isn't any air or liquid water there.

The Moon can be used as a "cosmic scorecard" for impacts. The oldest parts of the Moon, the lighter highlands areas, have too many craters to count, because collisions happened so often in the early solar system. The impact rate was over a thousand times what it is today. But the dark maria, which are mainly smooth and flat, are not as old as the highlands.

The rocks returned by Apollo astronauts show that the dark maria were formed by molten rock flows less than 4 billion years ago. If we looked at the Moon just after these maria formed, they would be craterless. Every crater we see today on these maria represents an impact during the last 4 billion years. This makes the dark maria a scorecard for how many big crater events there have been in the past 4 billion years. Count the big craters in the maria, divide by 4 billion years, and you can get an idea of how frequently the maria were impacted. Earth gets hit by a big object at about the same rate. And Earth's cross section is about 64 times larger than the surface area of the maria, so we need to factor that in.

Can you count how many craters there are in the maria regions of the Moon? Multiply that by 64 and divide by 4 billion years. That's an estimate of how often Earth gets hit by a large object, on average.

Galileo mission spacecraft view of Earth and the Moon from space. Earth's diameter is 4 times bigger than the diameter of the Moon but 16 times the surface area.

Think Questions

1. What kinds of animals might have survived the period of reduced light?

2. If a huge impact like the Chicxulub impact were to happen today, what might be the result worldwide?

3. Look at the outer ring of the Chicxulub Crater in the picture on page 39. How destructive would that crater be in your area? Get a local map or online map, and draw the size of the crater in your area to get an idea of the size of this event.

4. How often does such a big object hit Earth?

Investigation 6: Craters **41**

Gene Shoemaker: Astrogeologist

Eugene "Gene" Merle Shoemaker (1928–1997) was a planetary scientist who specialized in meteor impacts and the role they have played in the solar system. His passion was astrogeology, and he dreamed of going to the Moon. Almost single-handedly he created the discipline of planetary geology.

He died in a car accident while searching for meteor craters in Australia. Shoemaker had become well known to the Australians. Wherever he went, he captivated people's imaginations with his enthusiasm for geology, love of the land, and warm personality.

Shoemaker's claim to fame was his pioneering research on the formation of impact craters on the Moon, Earth, and other planetary bodies. He also discovered numerous Earth-crossing asteroids and comets. Shoemaker, his wife Carolyn (1929–), and a colleague, David Levy (1948–), discovered Comet Shoemaker-Levy 9, whose pieces crashed into Jupiter in July 1994. Together, the Shoemakers were the leading discoverers of comets in this century and are credited with discovering more than 800 asteroids.

Shoemaker seems to have been a geologist from the day he was born in Los Angeles, California. He graduated from the California Institute of Technology in Pasadena at the age of 19 and completed a master's degree only a year later. He joined the US Geological Survey and began exploring for uranium deposits in Colorado and Utah in 1948. These studies brought him near many volcanic features named Hopi Buttes, and an impact crater named Barringer Crater (also known as Meteor Crater), on the Colorado Plateau in the western United States.

From 1957 to 1960, Shoemaker carried out pioneering work on the nature and origin of Barringer Crater (near Winslow, Arizona). At that time, Shoemaker and his colleague Edward C. T. Chao (1919–2008) discovered shock quartzite (coesite), a mineral created only during impacts, in the rubble at the bottom of Barringer Crater. This mineral provided geologists with a substance to look for when investigating structures that might be ancient impact craters elsewhere on Earth. Soon thereafter, Shoemaker found coesite in the Ries basin in Germany, confirming that it was a giant impact structure. Ries basin was the second confirmed Earth crater. The discovery of coesite provided the tool geologists needed to identify many more impact structures on Earth. Eventually this discovery led to the theory that catastrophic impacts might have caused mass extinctions over geological time.

A man of vision, Shoemaker believed geological studies would be extended into space. In his early career, he dreamed of being the first geologist to map the Moon. However, a health problem prevented Shoemaker from being the first astronaut geologist. Even so, he helped train the Apollo astronauts and sat beside Walter Cronkite (1916–2009) during the evening news and gave geological commentary during the historic Moon walks.

During the 1960s, Shoemaker led teams that investigated the structure and history of the Moon and developed methods of planetary geological mapping using telescope images of the Moon. He was involved in the Ranger and Surveyor Moon-probe programs, continued with the manned Apollo programs, and culminated his Moon studies in 1994 as science team leader for the Clementine project.

Barringer Crater

Investigation 6: *Craters* **43**

Shoemaker and his wife, Carolyn

Tribute design by Carolyn C. Porco

Shoemaker was a very highly respected scientist. The University of Arizona awarded him an honorary doctorate of science in 1984. In 1992, he received the National Medal of Science, the highest science honor given in the United States, from President George H.W. Bush.

From the time he was a teenager, Shoemaker wanted to go to the Moon. He told friends that his inability to qualify for astronaut training was his greatest personal regret. Shortly before Shoemaker died, he said, "Not going to the Moon and banging on it with my own hammer has been the biggest disappointment in life."

In death, Shoemaker realized his life's great ambition. His colleagues and friends acknowledged his passion to travel to the Moon. They placed a small polycarbonate capsule carrying an ounce of Shoemaker's cremated remains in the *Lunar Prospector* spacecraft. The capsule, 4.5 centimeters (cm) long and 1.8 cm in diameter, was carried in a vacuum-sealed, flight-tested aluminum sleeve mounted deep inside the spacecraft.

The capsule was wrapped with a piece of brass foil inscribed with an image of Comet Hale-Bopp, an image of Barringer Crater in Arizona, and a passage from Shakespeare's enduring love story *Romeo and Juliet*.

When the tiny *Lunar Prospector* successfully crash-landed in a dark crater near the Moon's south pole on Saturday, August 31, 1999, it left the ashes of the pioneering astrogeologist on the lunar surface. There could be no finer tribute to the legendary planetary geologist who said his greatest unfulfilled dream was to go to the Moon.

The Cosmos in a Nutshell

Here on Earth we measure distance in meters (m) and kilometers (km). These units are way too short for measuring distance in the solar system and the **cosmos**. We measure these huge distances in **astronomical units** and **light-years**.

One astronomical unit (AU) is the average distance between Earth and the Sun, about 150 million km.

One light-year (ly) is the distance light travels in 1 year, at the speed of about 300,000 km per second. One ly is about equal to 9.5 trillion km, which is equal to about 63,000 AU. The closest star to us, other than the Sun, is about 4 ly away.

This means if we're recording a distance in astronomical units, it's a relatively small unit of measurement compared to a light-year.

We can use the terms *light-second*, *light-minute*, and *light-hour* as subunits for measuring distance.

One light-second is the distance light travels in a second. To talk to an astronaut on the Moon, we use radio waves, which also travel at the speed of light. A radio signal takes about 1.3 seconds to go from Earth to the Moon, so the Moon is about 1.3 light-seconds away.

One light-minute is the distance light travels in a minute. It takes about 8.5 light-minutes, the time it takes to eat an apple, for light to get to us from the Sun. We can say the Sun is a bit over 8 light-minutes away. To send radio signals to spacecraft on or near Mars can take as little as 5 minutes or as long as 20 minutes. The distance between Earth and Mars varies from about 5 to 20 light-minutes, depending on where the two planets are in their orbits.

One light-hour is the distance light travels in an hour. Sending radio waves to spacecraft visiting the outer planets of the solar system takes hours. It takes at least 1 hour and 20 minutes for radio waves to get to Saturn, and sometimes more than 4 hours to get to Neptune.

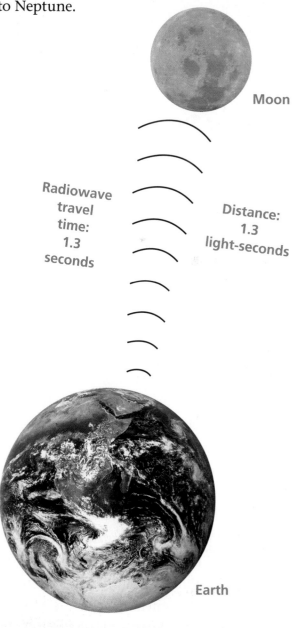

Moon

Radiowave travel time: 1.3 seconds

Distance: 1.3 light-seconds

Earth

Investigation 7: Beyond the Moon **45**

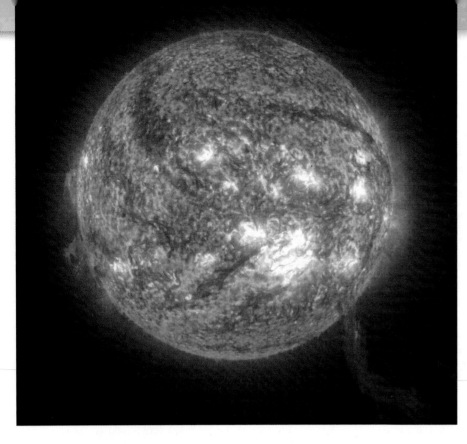

Because of its strong gravitational force, the Sun maintains the orbits of the planets.

The Solar System

The solar system is a region of space occupied by the Sun and all things orbiting around it. In rough order of size, from largest to smallest, those orbiting objects are the major planets, their **satellites** (moons), asteroids, comets, meteoroids, and dust. The Sun accounts for about 99.8 percent of the mass of the solar system. The 0.2 percent that was not drawn into the Sun forms everything else in the solar system.

A planet is an object that orbits a star and is massive enough for its own gravity to force it into a spherical shape. The number of known planets in the solar system has changed over the years. In ancient times, Earth's status as a planet was not understood. Only the five brightest planets (Mercury, Venus, Mars, Jupiter, and Saturn) were known and observed. But few, if any, people thought these planets orbited the Sun. It was not until telescopes were invented that other planets were discovered: Uranus in 1781 by an Englishman, William Herschel (1738–1822); Neptune in 1846 by a German, Johann Gottfried Galle (1812–1910), with orbit position calculations by an Englishman, John Couch Adams (1819–1892) and a Frenchman, Urbain Jean Joseph Leverrier (1811–1877); and Pluto in 1930 by an American, Clyde Tombaugh (1906–1997).

Saturn

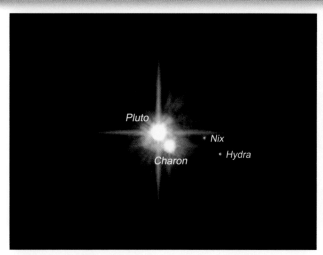

Three of Pluto's moons are shown here. A fourth moon of the dwarf planet was confirmed in July 2011.

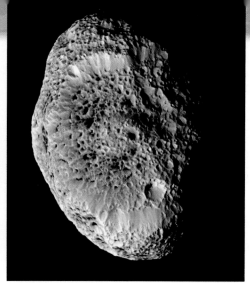

Hyperion, moon of Saturn

With more Pluto-like objects being found in the outer reaches of the solar system, members of the International Astronomical Union (IAU) had to make a tough decision about whether to call the new objects planets or to put them into a new category. The debate was quite intense. Some members called for a definition of *planet* based mostly on a minimum size. Others wanted to include the idea that a planet's gravity had to clear out all the other objects in the neighborhood of its orbit. That means any small local objects would be drawn into and merged with the planet. The **dwarf planet** was the new category that fit Pluto. A dwarf planet is an object that orbits the Sun and is big enough to be round, but doesn't clear the neighborhood of objects near its orbit. In 2008, the IAU again changed Pluto's classification to a **plutoid**. A plutoid is a type of dwarf planet.

A satellite is an object orbiting a larger object. A nonartificial object orbiting a planet is called a natural satellite, or moon, of that planet. We usually distinguish Earth's Moon from other moons by capitalizing the word, because it is also the name of Earth's natural satellite. Probes we launch into orbit around Earth or other planets are called artificial satellites.

An asteroid is a small, rocky object that orbits the Sun. Most asteroids orbit between the orbits of Mars and Jupiter. But many asteroids have orbits that take them closer to the Sun than Earth's orbit does. Others have orbits that take them well beyond Jupiter. The largest asteroid between Jupiter and Mars is named Ceres. It's about as wide as the state of Texas and is large enough to be called a dwarf planet under the current IAU definition.

An asteroid

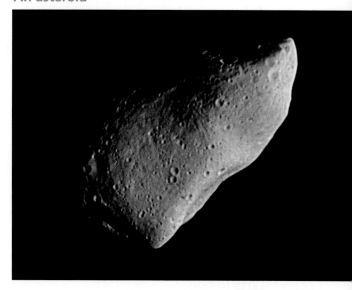

Investigation 7: Beyond the Moon

A meteoroid is a bit of solid debris in space. A small meteoroid (dust-sized to pebble-sized) enters Earth's atmosphere at an average speed of about 40,000 km per hour and vaporizes due to friction with the air. That can leave a streak of light across the night sky, which is called a meteor. Meteors are also called shooting stars, although they are nothing like stars in reality. If a meteoroid survives its fiery passage through the atmosphere, the pieces that hit the ground are called meteorites.

> *Meteoroid:* a bit of solid debris in space
>
> *Meteor:* a streak of light in the night sky, caused by a meteoroid vaporizing after entering Earth's atmosphere. A meteor is also called a shooting star.
>
> *Meteorite:* a meteoroid that has hit the ground

Beyond the orbit of Neptune is a region of the solar system called the **Kuiper Belt** (pronounced KI•per belt). The Kuiper Belt is populated by millions of small bodies called Kuiper Belt Objects (KBOs). While asteroids in the asteroid belt between Mars and Jupiter are mostly made of rock and metal, KBOs are made of ices, not only frozen water, but also frozen methane or ammonia. Over 1,000 KBOs have been discovered and cataloged. The Kuiper Belt extends from about the orbit of Neptune (about 30 AU from the Sun) out to about 55 AU from the Sun.

A comet is a chunk of ice, dust, and rock a few kilometers in size. When a comet comes close enough to the Sun, it can develop one or two "tails." The tails are made of gas and dust burned off the comet's surface by the Sun's energy as the surface material of the comet starts to boil.

This image of Comet Tempel 1 was taken 67 seconds after the *Deep Impact* spacecraft deliberately smashed into the comet to find out more about what comets are.

A comet's tail always points away from the Sun, no matter what direction the comet is moving. That's because light and particles coming from the Sun form solar wind and disperse the material boiled off the comet's surface. Comets spend most of their time very far from the Sun. They are active, with a visible tail, for only a few weeks or months as they swing quickly around the Sun in very long, oval orbits. Beyond the Kuiper Belt, astronomers think there might be a huge sphere of icy matter that could be a couple of light-years in diameter. This region, called the Oort cloud, could be the origin of comets.

The above objects are all part of the solar system. Review your notebook entry *Looking at the Cosmos*, and make any necessary changes to your solar system list.

Nebulae

A **nebula** is a cloud of gas and dust in space between stars. *Nebula* (plural nebulae) is Latin for cloud. Some nebulae glow with their own light. Some nebulae scatter light from stars within them, and some block light from things behind them.

Another way to sort nebulae is by how they relate to the evolution and life spans of stars. Many nebulae are where new stars are forming. In fact, a single nebula can be so huge that it can be the birthplace for a whole cluster of stars. But a nebula can also be leftover dust and debris after a star dies. When a large star dies, it can leave behind a supernova remnant, such as the Crab Nebula.

Stars

The Sun is an average-size star. Stars are large, hot balls of gas. They generate energy in their cores by nuclear reactions. Most stars are very stable. Depending on size, they can have life spans ranging from a few hundred thousand years to several billion or even trillions of years for smaller, slow-burning dwarf stars. Stars radiate in different colors, depending on how hot they are. Red stars are the coolest (still over 2,000°C on the surface), and blue stars are the hottest (over 10,000°C on the surface).

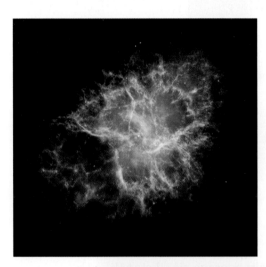

The Crab Nebula (NGC 1952)

The Eagle Nebula (M16)

Investigation 7: Beyond the Moon **49**

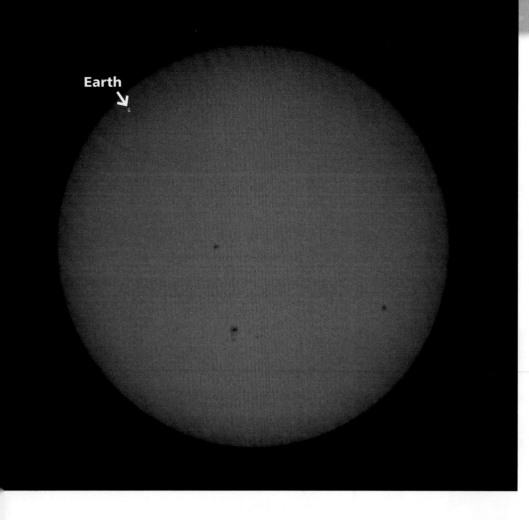

Current Earth/Sun ratio. In 5 billion years or so, Earth could be engulfed by the Sun if it becomes a red giant.

A system of two stars orbiting each other is called a binary star. Some systems have even more stars—triple or higher multiple-star systems. About half of all stars are members of multiple-star systems. Alpha Centauri, the nearest star to us other than the Sun, is a triple-star system with two stars similar to our Sun and one small, red star. They all orbit around one another. Very large, massive stars burn fuel much faster than smaller stars. Their life span may last only a few hundred thousand years. Smaller stars will live on for billions of years because the nuclear reactions take place much more slowly. Eventually, the star's fuel will begin to run out.

Many average-sized stars in the final stages of their lives can become red giants. A red giant is a very large star but with a relatively cool surface temperature. If the Sun becomes a red giant, in perhaps 5 billion years, its surface might extend past the orbit of Mars.

A relatively average-sized star (like the Sun) can shrink and become a white dwarf. This happens after its red-giant phase and after it has exhausted the fuel for its nuclear reactions. White dwarfs shine only by radiating away their stored-up heat. A white dwarf as massive as the Sun may shrink to the size of Earth, or 1/100 of its original size. It can become so compressed and dense that a teaspoonful of its material would weigh as much as a truck.

In comparison, at the end of the life span of a very large star, the star can become a **supernova**—a huge explosion that can generate more light than all the other stars in the **galaxy** combined. A supernova may shine brightly for several days.

Once a large star goes supernova, the remains at the core get crushed into a very dense object. This object can be either a neutron star or a **black hole**. For stars that start out 1.3 to 2 times as massive as the Sun, pressures in the supernova are so great that the electrons of atoms are forced into the atoms' nuclei. There they combine with protons and create all neutrons. The entire mass of this neutron star is crammed into a ball only about 10 km across. A teaspoonful of material in a neutron star would weigh more than all the automobiles in the United States put together.

For more massive stars, a black hole can form after the star goes supernova. Black holes are so dense that even the neutrons are crushed. The gravitational pull is so strong that nothing can escape from inside the black hole, not even light.

Star Clusters

A **star cluster** is a group of stars held together by their mutual gravitational attraction. One type of star cluster, called an open star cluster, has only a few dozen to several hundred stars. Open star clusters are sometimes found in a nebula, indicating that the stars have just formed.

A globular star cluster is spherical in shape, has older stars, and is larger than open star clusters. A globular star cluster can contain hundreds of thousands to a few million stars. About 150 globular star clusters have been found in the **Milky Way** galaxy. They form a huge halo around the galaxy's disk.

An open star cluster

A globular star cluster

A double star cluster

Galaxies

A galaxy is an enormous collection of tens of millions to hundreds of billions of stars, interstellar gas, and dust. A galaxy is held together by the gravity of its stars, which revolve around the center of the galaxy. There is good evidence that at the center of many large galaxies is a supermassive black hole with a mass equal to that of millions of stars.

The most common type of galaxy we see has a flattened, spiral shape called a spiral galaxy. Another type of galaxy, an elliptical galaxy, has an oval-like shape with no spiral patterns. And a few galaxies are irregular in shape.

From Earth, we see part of the Milky Way as a faint band of hazy light stretching all the way across the sky from north to south. It can be seen only from clear, dark locations. Through binoculars or a small telescope, we can see that it is made of vast numbers of faint stars. This is actually the disk of our own spiral galaxy as seen from our point of view out toward the edge of the disk.

The Milky Way is a spiral galaxy.

This image shows the elliptical galaxy NGC 4636 and the hot gas extending 25,000 ly.

Galaxies M81 (left), a spiral galaxy, and M82 (right), an irregular galaxy, are about 12 million ly away from us.

A closer look at the M81 galaxy

The Milky Way is about 100,000 ly across and contains roughly 400 billion stars. The Sun is about two-thirds of the way out from the center of the galaxy. The Sun completes one counter-clockwise orbit around the center of the Milky Way about every 200 million years. The center of our galaxy is believed to be a black hole. The closest galaxies to the Milky Way are the Magellanic Clouds. These galaxies are about 100,000 ly away and can be seen easily from Earth's Southern Hemisphere.

The Milky Way is a member of a relatively small cluster of about two dozen galaxies called the Local Group. Most of the Local Group galaxies are considerably smaller than the Milky Way. The only other large galaxy is the Andromeda Galaxy, which is about 2 million ly away. The Local Group is about 3 million ly across and is itself part of a supercluster of galaxy clusters called the Virgo Cluster. The Virgo Cluster is named because its center is in the direction of the constellation Virgo.

Universe and *cosmos* are words used to describe all things that can be observed or detected. Astronomers' theory for how the **universe** came into being is known as the big bang. Many astronomical observations support the **big bang theory**, which suggests that there was one explosion from which clusters of galaxies formed and can now be seen moving apart from one another. Astronomers calculate that the big bang happened about 13.73 billion years ago.

Galaxy NGC 3628, a spiral galaxy seen edgewise, is 35 million ly away from Earth.

Galaxy cluster Abell S0740, an elliptical galaxy, as seen from Hubble Space Telescope, is about 450 million ly from Earth. It is comparable in size to the Milky Way.

 Take one last look at your notebook entry *Looking at the Cosmos*, and make any necessary changes to your lists for what is in the solar system, galaxy, and universe.

An artist's representation of the Milky Way galaxy and the location of the Sun

Investigation 7: Beyond the Moon **53**

How Earth Got and Held onto Its Moon

Counting out from the Sun, Earth is the first planet with a satellite, or moon. Mercury and Venus, nearer to the Sun, don't have a moon. Mars, the fourth planet out, has two moons, which are probably just a couple of asteroids that got caught by Mars and ended up in orbit after they were fully formed. It is suspected that Earth didn't have a moon at first, but acquired one early in its history as a result of a huge planetary collision. Visualize the event as it may have happened, perhaps 4.5 billion years ago.

Earth was pretty much formed as a planet. Most of the dust and gas in the region had been pulled in by the gravity of the proto-Earth. These were the early days of the solar system. If you had been there, you would have noticed a lot of material flying around in unstable orbits. Some of the chunks, called planetesimals, were the size of small planets themselves.

Planetary scientists now think that one of these planetesimals, perhaps the size of Mars, was traveling around the Sun in an orbit that was far from circular. It's not known why it had such an irregular orbit. Perhaps it was pulled by the gravity of a large planet, or perhaps there were lots of objects following irregular paths early in the solar system's history. Anyway, it ended up heading for Earth.

Had you been on Earth to see the event, here is what you might have observed. The planetesimal first appeared as a dot in the sky.

The Moon

Over a period of days and weeks, it grew bigger and bigger until it completely filled the field of view from Earth in that direction. Then it struck. Because the colliding objects were so large, the impact itself seemed to happen in slow motion. It lasted several minutes, even though the planetesimal was traveling at perhaps 40,000 kilometers (km) per hour.

The crash caused a chain of reactions. First, the impact destroyed the incoming object. The planetesimal was reduced to vapor, dust, and debris, with some pieces driven deep into the interior of Earth. A significant portion of Earth disintegrated as well. The energy that resulted from the crash produced an

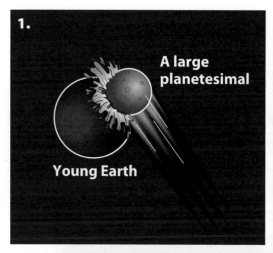
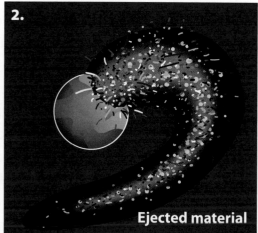
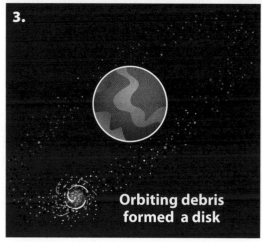
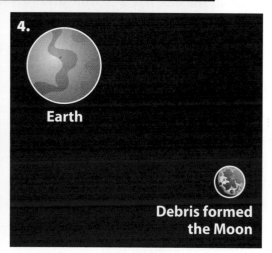

A representation of how the Moon formed

explosion of unimaginable magnitude. Earth itself might have been in danger of being blasted apart.

Second, the explosive release of energy threw a tremendous amount of matter into motion—at least 20 billion cubic kilometers (km³) of matter. Some of the most energized pieces of matter flew out into space and were never seen again. Other matter flew up into the air and then returned to Earth. Some returned almost immediately as huge rocks. Some returned a little later as granules of various sizes. And other matter returned months or even years later in the form of dust and chemicals held up in the atmosphere.

Finally, a significant portion of the debris didn't fly off into space, and it didn't return to Earth. It began orbiting Earth and formed a disk of debris, like the rings of Saturn. The ring was probably about two Earth diameters from the surface of Earth. Over a period of 1 or 2 years, the pieces of matter and dust started to attract one another. Gradually, they formed larger and larger chunks of debris, and eventually formed the Moon.

Earth had a moon where previously there was none. It must have been quite a sight up there only about 30,000 km above Earth, rather than the 384,000 km of today.

How is the Moon formation like solar system formation? Make a notebook entry explaining your ideas.

Investigation 7: Beyond the Moon **55**

The Role of Gravity

Gravity is one of the four known forces in the universe, along with electromagnetic force and two kinds of nuclear force. These four forces make everything in the world behave in ways we understand. The Law of Universal Gravitation tells us that gravity is the force that causes two masses to attract each other. The force of gravity exerted between two small masses, like a marble and an apple, is so small that we can't detect it. But the gravity exerted by a large mass, like a planet or a star, is tremendous. The larger the mass, the stronger the force of gravity it exerts.

The force of gravity between the Sun and Earth is so strong that it keeps Earth in a circular orbit around the Sun. The diagram below shows those forces in action. And it shows what would happen to Earth's path without the gravitational attraction.

When the planetesimal hit Earth, some matter was thrown straight up in the air. There were two possible outcomes for that matter. Chunks of debris thrown with enough velocity would escape Earth's gravity and become loose space debris. Debris without enough velocity would be pulled back down to the surface by Earth's gravity.

However, most of the matter ejected by the impact would not go straight up, but would be launched at an angle. Something different could happen to this matter.

Before we get into where that matter went, consider another piece of information about the behavior of matter. Sir Isaac Newton (1642–1727) figured out that an object in motion will travel in a straight line until it is acted on by a force that changes its direction. In other words, things don't travel in curves, circles, spirals, or zigzags, unless a force (a push or a pull) acts on them to change their motion.

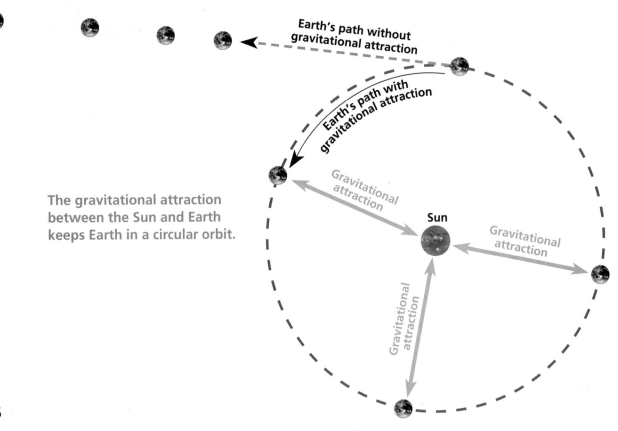

The gravitational attraction between the Sun and Earth keeps Earth in a circular orbit.

So where did the matter from the impact debris go? Most of the matter flew out from the impact at an angle, in a straight line but not straight up. Earth's gravity also acted on this debris. Take a rock the size of a microwave oven as an example. It flies off in a straight line, but at an angle. If there were no gravity, the rock would just keep going. But there is gravity, and it pulls on the rock, bringing it back to Earth. But the rock is going sideways so fast that the pull only changes the direction of the rock as it moves. When the planetesimal hit Earth, a lot of the material was thrown out at an angle just far enough to go into orbit around Earth. Earth's gravity caused it to orbit, and the debris became the Moon.

Imagine taking a yo-yo by the end of its string and swinging it around over your head. You have it going in a nice circle. If you let go of the string, what happens? It stops going in a circle and flies off in a straight line. As long as you keep applying a force (pulling on the string) to change the direction of the yo-yo, it continues to orbit your hand.

Gravity is the "string" pulling on the Moon to keep it in a circular path. Similarly, gravity is the force keeping Earth (and the other planets) in a circular orbit around the Sun. In fact, everything that is behaving in a predictable way in the solar system is orbiting something else. And in every case, gravity rules the action.

Tides are the twice-daily rising and falling of sea level. Tides also demonstrate the action of gravity, the influence of gravity of the Moon on Earth. Visit www.FOSSweb.com to view the "Tides" interactive and learn more about tides.

Go back to your notebook entry comparing the Moon formation and solar system formation and add some ideas about gravity. How did gravity affect each formation?

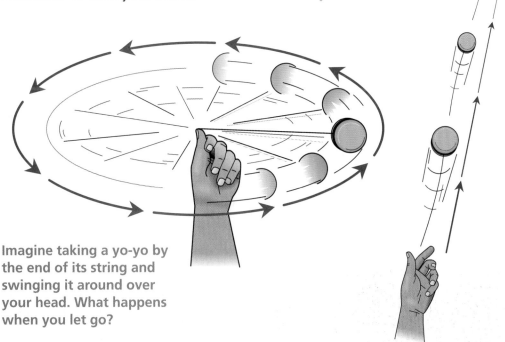

Imagine taking a yo-yo by the end of its string and swinging it around over your head. What happens when you let go?

Investigation 7: Beyond the Moon 57

Sizes and distances of solar-system objects are not drawn to scale.

A Tour of the Solar System

Imagine you are coming to the solar system as an alien stranger. A tour guide provides information as you gaze out the window. What will you see?

As you approach, you get a bird's-eye view from space. From here, you see the whole solar system. The most surprising thing is that the solar system is mostly empty. The matter is concentrated in tiny dots that are extremely far apart. Most of the dots are planets. In the same way that the Moon is held in orbit around Earth, the planets are held in orbit around the Sun by gravity.

There is a star in the center of the solar system. Four small planets orbit pretty close to the star. These are the rocky terrestrial planets.

Next, there is the asteroid belt, a region of small bits of matter orbiting the star.

Out farther, four big gas planets are in orbit around the star. These are the gas giant planets.

Beyond the gas giant planets is a huge region of different-size icy chunks of matter called the Kuiper Belt. Some of the chunks are big enough to be planets. Others have orbits that send them flying through the rest of the solar system. That's all that can be seen from your bird's-eye view in space.

The Sun

The Sun is a star. It is similar in size, color, and brightness to some of the stars you can see in the night sky. The Sun is at the center of the solar system. Everything else in the solar system orbits the Sun. The Sun rules.

The Sun is made mostly of hydrogen (74 percent) and helium (25 percent). It is huge. The diameter is about 1,384,000 kilometers (km). That's about 109 times the diameter of Earth. (See Earth compared to the Sun in the picture.)

The Sun is incredibly hot. Scientists have figured out that the temperature at the center of the Sun is 15,000,000 degrees Celsius (°C). The temperature of the Sun's surface is lower, about 5,500°C. Thermonuclear reactions in the Sun's core create heat and light energy. About 3.6 tons of the Sun's mass is being changed into heat and light energy every second. This energy radiates out from the Sun in all directions. A small amount of it reaches Earth.

Another name for the Sun is Sol. The solar system is named for the ruling star. The reason the Sun rules is its size. The Sun has 99.8 percent of the total mass of the solar system. All the other solar-system objects travel around the Sun in predictable almost-circular paths called orbits. The most obvious objects orbiting the Sun are the planets.

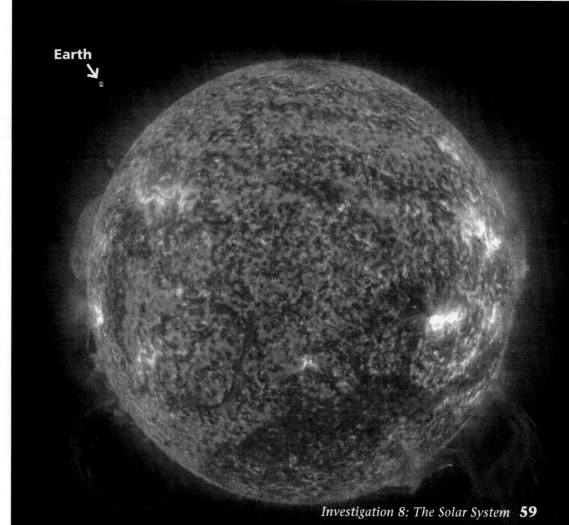

Earth orbits the Sun.

Investigation 8: The Solar System **59**

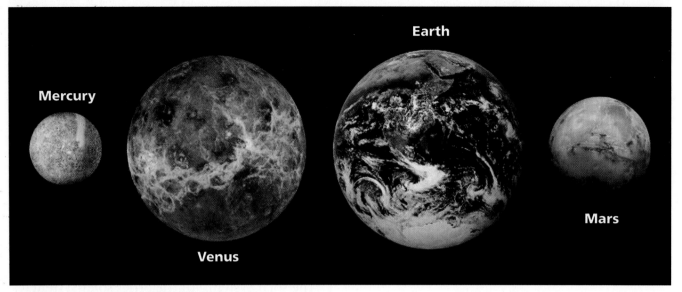

Relative sizes of the terrestrial planets

Terrestrial Planets

The terrestrial planets are the four planets closest to the Sun. The terrestrial planets are small and rocky.

Mercury

Mercury is the closest planet to the Sun. It is smaller than Earth and has no natural satellite. By human standards, it is an uninviting place. Mercury is very hot on the side facing the Sun and very cold on the side away from the Sun. It has no atmosphere or water.

Mercury is covered with craters. The craters are the result of thousands of collisions with objects flying through space. The surface of Mercury looks a lot like Earth's Moon.

Venus

Venus is the second planet from the Sun. It is about the same size as Earth and has no natural satellite. The surface of Venus is very

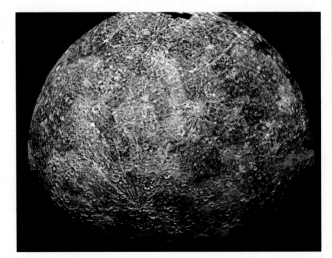

Mercury's surface is rough with craters.

The surface of Venus is dry and covered with volcanoes.

hot all the time. It is one of the hottest places in the solar system. It is hot enough to melt lead!

There is no liquid water on Venus. It has an atmosphere of carbon dioxide. The dense, cloudy atmosphere makes it impossible to see the planet's surface. Modern radar, however, allows scientists to take images through the clouds. We now know that the surface of Venus is dry, cracked, and has many volcanoes.

Earth

Earth is the third planet from the Sun. It has a moderate temperature all the time. It has an atmosphere of nitrogen and oxygen, and it has liquid water. As far as we know, Earth is the only place in the universe that has life. Earth also has one large satellite called the Moon, or Luna. The Moon orbits Earth in about a month. The Moon is responsible for the tides in Earth's oceans. The Moon is the only extraterrestrial place humans have visited.

Moon

Earth is 150 million km from the Sun. This is a huge distance. It's hard to imagine that distance, but think about this. Sit in one end zone of a football field and curl up into a ball. You are the Sun. A friend goes to the other end zone and holds up the eraser from a pencil. That's Earth. Get the idea? Earth is tiny, and it is a long distance from the Sun. Still, the solar energy that reaches Earth provides the right amount of energy for life as we know it.

Earth

Investigation 8: The Solar System

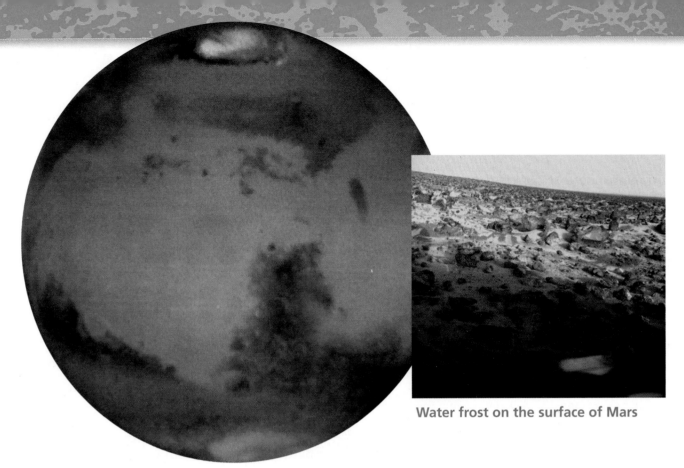

Water frost on the surface of Mars

Mars

Mars is the fourth planet from the Sun and has two small satellites, Phobos and Deimos. Mars is a little like Earth, except it is smaller, colder, and drier. There are some places on Mars that are somewhat like Death Valley in California. Other places on Mars are more like Antarctica and the volcanoes of Hawaii.

Mars is sometimes called the Red Planet because of its red soil. The soil contains iron oxide, or rust. The iron oxide in the soil tells scientists that Mars probably had liquid water at one time. But liquid water has not been on Mars for 3.5 billion years. It has frozen water in polar ice caps that grow and shrink with the seasons.

Mars is the next likely place humans will visit. But exploring Mars will not be easy. Humans can't breathe the thin atmosphere of carbon dioxide. Astronauts will need to wear life-support space suits for protection against the cold.

A drawing of the Phoenix Mars lander

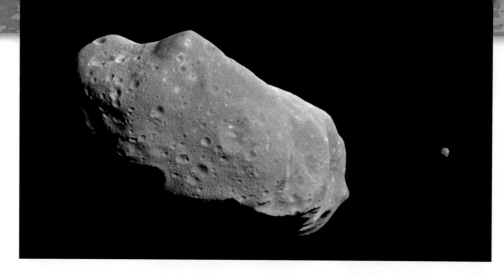

Asteroid Ida with satellite Dactyl

Several robotic landers have explored Mars and sent back information about the surface. In 2008, the Phoenix Mars lander confirmed the presence of ice and water vapor in the soil. The Mars Reconnaissance Orbiter has returned images of a constantly changing Martian surface.

Asteroids

Beyond the orbit of Mars, there are millions of chunks of rock and iron called asteroids. They all orbit the Sun in a region called the asteroid belt. The asteroid belt is like the boundary of the terrestrial planets. When the spacecraft *Galileo* flew past asteroid Ida in 1993, scientists were surprised to find that it had a satellite. They named the tiny moon Dactyl. The biggest asteroid is Ceres. It is about 960 km in diameter. Ceres is also called a dwarf planet.

Gas Giant Planets

The next four planets are the gas giant planets, which are made mostly of gas. They do not have rocky surfaces like the terrestrial planets. So there is no place to land or walk around on them. They are much bigger than the terrestrial planets. What we have learned about the gas giant planets has come from probes sent out to fly by and orbit them. Even though the gas giant planets are all made of gas, each one is different.

Relative sizes of the gas giant planets

Investigation 8: The Solar System

Jupiter

Four of Jupiter's satellites: Io, Europa, Ganymede, and Callisto

Jupiter

Jupiter is the fifth planet from the Sun. It is the largest planet in the solar system. It is 11 times larger in diameter than Earth. Scientists have found evidence of at least 63 moons orbiting Jupiter. The four largest moons, Io, Europa, Ganymede, and Callisto were first described by Galileo Galilei (1564–1642) in 1610.

Jupiter's atmosphere is cold and poisonous. It is mostly hydrogen and helium. The stripes and swirls on Jupiter's surface are cold, windy clouds of ammonia and water. Its Great Red Spot is a giant storm as wide as three Earths. This storm has been going on for hundreds of years. On Jupiter, the atmospheric pressure is so strong it squishes gas into liquid. Jupiter's atmosphere could crush a metal spaceship like a paper cup.

Saturn

Saturn is the sixth planet from the Sun. It is the second largest planet and is very cold. There is evidence of at least 62 moons orbiting Saturn. Saturn is made up mostly of hydrogen, helium, and methane. It doesn't have a solid surface. It has clouds and storms like Jupiter, but they are harder to see because they move so fast. Winds in Saturn's upper atmosphere reach 1,825 km per hour.

The most dramatic feature of Saturn is its ring system. The largest ring reaches out 200,000 km from Saturn's surface. The rings are made of billions of small chunks of ice and rock that are spaced fairly far apart. All the gas giant planets have rings, but the others are not as spectacular as Saturn's.

False-color image of Saturn

Close-up of the rings of Saturn

Uranus and three of its moons: Miranda, Titania, and Oberon

Uranus

Uranus is the seventh planet from the Sun. There is evidence of at least 27 moons and 11 rings orbiting Uranus. Uranus is very cold and windy, and would be poisonous to humans. It is smaller and colder than Saturn.

Uranus has clouds that are extremely cold at the top. Below the cloud tops, there is a layer of water, ammonia, and methane. Like other gas giant planets, Uranus may be very hot at its core. Uranus appears blue because of the methane gas in its atmosphere.

Neptune

Neptune is the eighth planet from the Sun. There is evidence of at least 13 moons and 4 thin rings orbiting Neptune. Neptune is the smallest of the gas giant planets, but is still the fourth largest planet in the solar system.

Neptune is made mostly of hydrogen and helium with some methane. It might be the windiest planet in the solar system. Winds rip through the clouds at more than 2,000 km per hour. Scientists think there might be an ocean of extremely hot water under Neptune's cold clouds. It does not boil away because of the incredible pressure on the planet.

Neptune and one of its moons, Proteus

Pluto and one of its moons, Charon

Pluto

Out beyond the gas giant planets is a disk-shaped zone of icy objects called the Kuiper Belt. Pluto is one of the Kuiper Belt objects. Some scientists considered Pluto a planet because it is massive enough for its gravity to hold it in the shape of a sphere. Others did not consider Pluto a planet. To them, Pluto was just one of the large pieces of debris in the Kuiper Belt. Scientists now classify Pluto as a plutoid, a type of dwarf planet.

Pluto has a thin atmosphere. It is so cold that the atmosphere actually freezes and falls to Pluto's surface when it is farthest from the Sun. Even though Pluto is smaller than Earth's Moon, it has its own satellites. They are named Charon, Nix, and Hydra.

Eris

In July 2005, astronomers at the California Institute of Technology announced the discovery of a new planet-like object. It is called Eris. Like Pluto, Eris is a Kuiper Belt object and a plutoid. But Eris is more than twice as far away from the Sun as Pluto is! The picture gives an artist's idea of what the Sun would look like from a position close to Eris.

A painting showing that the Sun would look like a bright star from Eris

Comets

Sometimes comets are compared to dirty snowballs. Scientists think comets might provide valuable information about the origins of the solar system.

Comets orbit the Sun in long, oval paths. Most of them travel far beyond the orbit of Pluto. A comet's trip around the Sun can take hundreds or even millions of years, depending on its orbit. A comet's tail shows up as it nears the Sun and begins to warm. The gases and dust that form the comet's tail always point away from the Sun.

Comets' orbits can cross those of the planets. In July 1994, a large comet, named Comet Shoemaker-Levy 9, was on a collision

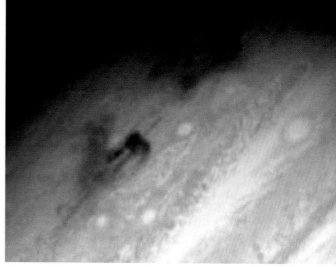

Two of the 21 larger than Earth-size craters on Jupiter

course with Jupiter. As it got close to Jupiter, the comet broke into 21 pieces.

The pieces slammed into Jupiter for a week. Each impact created a crater larger than Earth in Jupiter's surface.

The Stardust Mission collected particles from Comet Wild 2 and returned them to Earth in 2006. Scientists will continue to study these particles for many years to try to understand more about the compositions and origins of comets.

Comets have been called dirty snowballs.

Comet Shoemaker-Levy 9 broke into pieces as it got closer to Jupiter.

Think Questions

1. What is the Sun, and what is it made of?
2. What is the solar system?
3. Which planets are terrestrial planets? Which planets are gas giant planets?
4. What is found in the Kuiper Belt?
5. Which planet has the most moons orbiting it?
6. How are asteroids and comets alike and different?

Hunt for Water Using Spectra

Have you used a **spectroscope** to look at the spectrum of colors coming from sources of lights around you? If so, you might recall seeing rainbows like those in the incandescent-light spectrum and fluorescent-light spectrum on this page. Different patterns of color reveal the identity of elements or compounds that are **emitting** the light.

Incandescent light shows a mostly even spread of rainbow colors that is characteristic of the tungsten metal glowing in the white-hot lightbulb filament. The Sun's spectrum looks very much like this, but is very bright.

Fluorescent light shows several very bright color lines standing out against a dimmer spectrum. The main bright lines are given off by the element mercury.

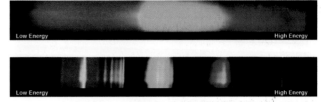

Incandescent-light spectrum

Fluorescent-light spectrum

Imagine you are on a NASA Mission to find a place in the solar system that has water, one of the essential necessities for life. The problem you face is that you can't go there to get samples.

If you want to study faraway places, like moons, planets, and stars, you need to get information from the light they give off or reflect. Two very simple properties of light can provide powerful information. They are the light's brightness and its color.

Electromagnetic Spectrum

Light energy comes to us in waves. Light acts like vibrating electric and magnetic force fields. We can measure the light's wavelength (how long the waves are) and energy. Colors with longer wavelengths have lower energy. Colors with shorter wavelengths have higher energy.

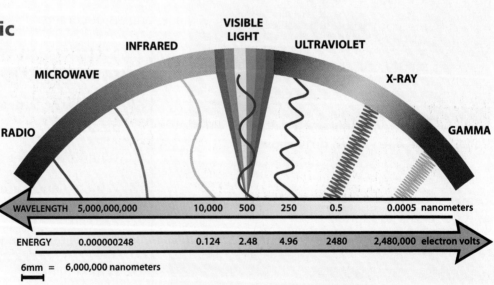

Colors of visible light range from red (low energy, long wavelengths) to violet (high energy, short wavelengths). Higher energies than violet, invisible to our eyes, include ultraviolet light (UV), X-rays, and gamma rays. Lower energies than red, also invisible to our eyes, include infrared light (IR), microwaves, and radio waves.

Fingerprinting Matter

Every element and compound is unique and has a unique spectral signature. Much like fingerprinting, the spectral signature can be used to identify elements and compounds.

Here is the spectrum of hydrogen, one of the two elements in water molecules. We can see two kinds of lines in its spectrum.

Emission Lines. Hydrogen atoms emit (give off) light of specific wavelengths or frequencies. Bright lines show the wavelength of the emitted light.

Hydrogen emission spectrum

Absorption Lines. Hydrogen atoms also absorb light in related wavelengths and frequencies. Dark areas in the image above (most of the spectrum) show that the light at those wavelengths is being absorbed, so it is not emitted or seen.

Seeing the Invisible

Human eyes can see wavelengths only of certain sizes. But hydrogen emits "light" at wavelengths that we cannot see with our eyes, or even with the spectroscope that we used in class. These wavelengths are in the ultraviolet region of the **electromagnetic spectrum**, invisible to us. However, scientists can use tools to detect the wavelengths that our eyes cannot see. For example, a space probe could detect gamma rays in a part of the spectrum where there is no visible light. Even though we wouldn't see anything with our eyes (or even with a telescope or classroom spectroscope), other scientific tools designed to detect gamma rays could be used to observe a powerful energy event like a supernova.

 What part of the electromagnetic spectrum were you able to analyze using the spectroscope? Make a notebook entry.

Spectral Signature of Water

Hydrogen and oxygen are the atoms that make water. But when we look at water vapor (gas) with a spectroscope, we see many more lines than either the spectrum of oxygen or hydrogen have separately. This is because the elements hydrogen and oxygen have formed a compound called water. We do see strong hydrogen lines in the spectrum, and if we look carefully, we can probably pick out oxygen lines.

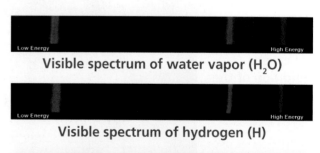

Visible spectrum of water vapor (H_2O)

Visible spectrum of hydrogen (H)

Visible spectrum of oxygen (O_2)

Investigation 9: Space Exploration

However, many of the important wavelengths for the spectrum of water are outside the visible spectrum, in the infrared and microwave ranges. More powerful tools let scientists detect these wavelengths and confirm the presence of water.

Water on Jupiter

Part of the NASA Juno Mission to Jupiter is to search for water. *Juno* launched in August 2011 and is scheduled to arrive at Jupiter in 2016.

Scientists debate over the origin and early development of the solar system, and they need more information. Jupiter, largest of the planets, holds answers to critical questions about the formation of the solar system. The primary scientific goal of the Juno Mission is to improve our understanding of Jupiter's structure, formation, and evolution.

One of *Juno*'s instruments is like a spectrometer but is designed to detect energy emissions in the microwave region of the electromagnetic spectrum. Yes, that's the same kind of energy we use in microwave ovens. The instrument is called a microwave radiometer and is used to measure the abundance of hydrogen in Jupiter's atmosphere. From this observation scientists can infer the amount of water in Jupiter's atmosphere.

The Juno Mission hopes to provide answers to science questions to reveal how the solar system formed. The origins of life itself may have critical ties to the special conditions under which planetary systems such as the Milky Way galaxy or solar system were created and evolved. Using data from this mission to Jupiter, scientists will come closer to understanding these conditions and their connections to the origins of the human species.

For more information about the Juno Mission, go to: www.FOSSweb.com.

An artist's rendition of the *Juno* spacecraft

Finding Planets outside the Solar System

Finding a planet circling a star other than the Sun is not easy. Such planets are called **exoplanets**. Finding them presents two problems.

Problem 1. Other stars are just too far away. Suppose a star is only 33 light-years (ly) away. That's a fairly close neighboring star, but it's about 330,000,000,000,000 kilometers (km) away. Looking for a planet the size of Earth around a star that far away is like trying to see a person's nose 20 km away. Our most powerful telescope, the Hubble Space Telescope, might do the job except for Problem 2.

The Hubble Space Telescope

Problem 2. The light coming from the star is just too bright. Did you ever try to see an airplane or bird flying by in the direction of the Sun? You should never look directly at the Sun, but you wouldn't be able to see objects flying in the direction of the Sun anyway. The intense light from the Sun would force you to look away. Stars put out a tremendous amount of light. The glare from a star overpowers any light reflected from the planet. A planet, which produces no light of its own, would be impossible to see in the glare of its star.

Because of the tiny size of planets at such great distances, and the glare of the star, it is not possible to find planets close to their stars by looking for them directly. To date, the two main planet-searching strategies use evidence of stars that move back and forth slightly and evidence of stars that get slightly dimmer. For now, let's call them the wobble method and the transit method.

Wobble Method

Scientists know that the pull of gravity keeps planets in orbit as they revolve around their star. This first figure is a simple diagram of a planet orbiting a star.

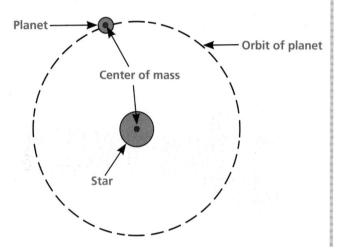

But closer analysis of a planet and star system shows that the star's gravity pulls on the planet *and* the planet's gravity pulls on the star. As a result, the center of mass of the star is not the point around which the planet orbits. This second diagram shows how both the planet and the star orbit their common "center of gravity," called the barycenter. The star's orbit is tiny, but the planet's orbit is large.

How much off-center is the star? If the solar system consisted of only the Sun and Jupiter, the barycenter would be just outside the Sun's surface.

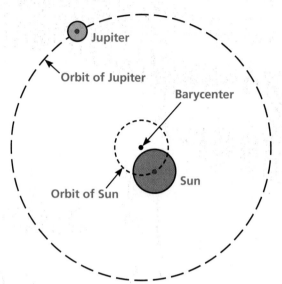

If the solar system had only the Sun and Earth, the barycenter would be deep within the Sun, but not at the center of the Sun. A larger planet would cause a bigger wobble (back-and-forth motion) of the star, because larger planets have more gravitational pull.

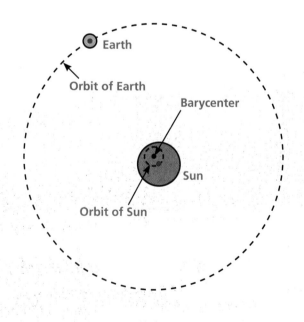

We can detect a planet indirectly by looking at a star's spectrum, the colors of light it gives off, for evidence of a wobble. This is spectroscopic planet searching. It works because astronomers can observe the spectrum lines produced by specific gases in a star (helium, hydrogen, and others). Spectrum lines correspond to wavelengths of light (specific colors of the rainbow) emitted by the gases.

The spectrum lines shift slightly if the star is moving toward or away from the observer. The lines shift toward the violet end of the spectrum if the star is moving closer to us. The lines shift toward the red end if the star is moving farther away. This is known as the Doppler shift. The Doppler shift in light waves is similar to the effect you hear as a siren, a loud car engine, or a train passes by. The pitch of the siren is higher when the sound source is moving toward you and lower when the sound source is moving away from you. In a sense, the sound waves get pushed together when they're moving toward you and stretched apart as they move away from you.

In the fall of 1996, a team of Swiss astronomers, including Michel Mayor (1942–) and Didier Queloz (1966–), announced that they had detected a Jupiter-mass object orbiting a nearby Sun-like star called 51 Pegasi.

Astronomers Geoffrey Marcy and Debra Fischer

They used the wobble method to detect the planet. By 2006, a team of astronomers at University of California, Berkeley, including astronomers Geoffrey Marcy (1954–), R. Paul Butler (1960–), and Debra Fischer (1953–), used the wobble method to discover over 170 more possible planet candidates. As of January 2012, over 700 exoplanets had been discovered and confirmed. Go to FOSSweb for updates on this number.

What we have referred to as the wobble method of finding exoplanets is more often called the spectroscopic method of planet finding. That's because it is really the Doppler shift of spectral lines of the star that provides evidence of the star's wobble. And the star's wobble is evidence of an exoplanet's gravitational effect on its star.

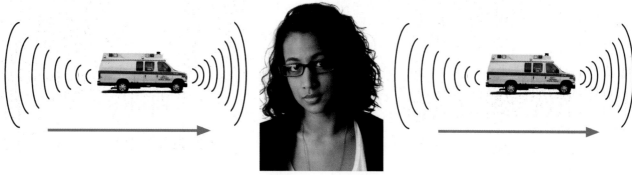

Investigation 10: Orbits and New Worlds

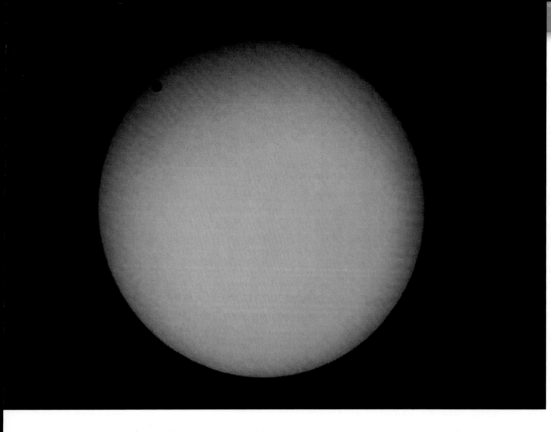

Transit of Venus across the Sun

Transit Method

If a planet passes directly in front of its star, the event is called a **transit**. During a transit, the planet blocks a small amount of the star's light, causing a slight dimming of the star called a blink. The bigger the planet is, the bigger the drop in brightness, and the bigger the blink. Even for the largest planets, it's not much of a blink. An observer from another planetary system would see a 1 percent drop in the Sun's brightness when the largest planet, Jupiter, passes in front of the Sun. That's 1 part in 100. Earth would cause only a 0.01 percent drop in brightness, or 1 part in 10,000. The "blink" would actually be a long drop in brightness—2 to 16 hours—but short compared with the time it takes the planet to orbit its star.

Observing star blinks is the transit method of detecting exoplanets. As of October 2008, about 52 planets had been discovered by the transit method. Most were found by astronomers using ground-based observatories. But four planets were discovered using a special brightness sensing satellite called CoRoT (Convection, Rotation, and planetary Transits). CoRoT was intended to probe the inner structure of stars by sensing tiny changes in star brightness, but it can also detect when a planet passes in front of its star. The Kepler Mission launched in March 2009. In February 2011, NASA released data of over 1,200 exoplanets candidates detected by the transit method using the Kepler space telescope.

For centuries, astronomers debated the probability of other planets existing in the universe. Now that debate can be abandoned.

For an up-to-date count of how many exoplanets have been discovered, visit www.FOSSweb.com.

Kepler Mission

The discovery of planets elsewhere in the universe raises an even more interesting question: Is anything living on any of those planets?

NASA's Kepler Mission is the first mission capable of finding Earth-sized and smaller planets around other stars. The hundreds of previous planet discoveries were of three types: gas giant planets, hot super-Earths in short-period orbits, and ice giants. Kepler uses the transit method to find small planets (one-half to twice the size of Earth), especially those in the habitable zone of their stars where liquid water (and possibly life) might exist. The Kepler Mission is specifically designed to survey a small region of the Milky Way galaxy to find Earth-sized planets in the habitable zone. The results will allow us to estimate how many of the billions of stars in the Milky Way have such planets with the potential for life.

Kepler scientists will be observing about 100,000 stars for at least 3.5 years. They will be collecting data about the *period* of the brightness drop, the *duration* of the brightness drop, and the *consistency* of the change in brightness, to confirm that a planet has been discovered. Once a new planet is confirmed, the Kepler science team can use the observation data and Kepler's third law of planetary motion to estimate the radius of the planet's orbit from the period (how long it takes the planet to orbit the star) and the mass of the star. The size of the planet is determined from the brightness drop of the star and the size of the star. From the **orbital period** and **orbit radius** and the temperature of the star, the planet's probable temperature can be calculated. Finally, the question of whether the planet is habitable (not necessarily inhabited) can be answered.

An artist's representation of the *Kepler* spacecraft in space

Life in the Universe

The universe is a large study site. Scientists looking for planets focus their attention on stars within a few dozen to 3,000 ly from Earth. In the Milky Way, however, there are 200 billion or more stars distributed in a spinning disk that is about 100,000 ly across and 10,000 ly thick. The vast majority of the stars in our galaxy are too far away to search at this time.

But it doesn't end there. The universe is populated with countless other galaxies, each with billions of stars. Some of the galaxies are like ours in structure—a disk with starry arms reaching out from a central bulge. Others are elliptical or irregular in shape. The distances between galaxies, even those in the Local Group, are so great that searching for planets outside the Milky Way is not possible at this time.

But knowing that planets do exist around other stars makes us dream of other worlds like Earth, maybe in our galaxy, and maybe beyond. And somewhere out there, at this very moment, there may be students in classrooms on those planets wondering . . . is there life anywhere else in the universe?

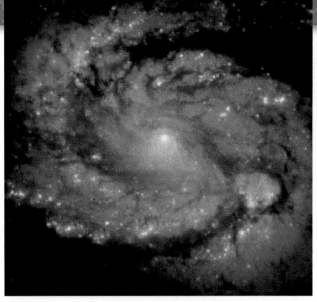
Spiral galaxy M100 is similar to the Milky Way.

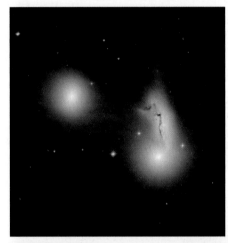
A spiral galaxy between two elliptical galaxies

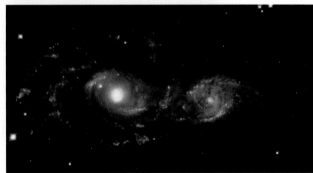
Spiral galaxy NGC 4639 is 78 million ly away.

Spiral galaxy NGC 4414 is 60 million ly away.

Think Questions

1. Once we confirm other planetary systems, what do you think the next stage of exploration should be?
2. Why don't we have a photo of the entire Milky Way like those of other galaxies seen on this page?

Images and Data

Images and Data Table of Contents

Investigation 1: Where Am I?
White House **79**
White House Neighborhood **80**
White House Community **81**
Washington, DC, Area **82**
Northeast Region **83**
Earth **84**

Investigation 3: Seasons
Worldwide Sunrise/Sunset Data **85**

Investigation 4: Moon Study
Full Moon **86**
Moon Image **87**
Earth/Moon Comparison **88**
Sun, Planets, and Satellites in the
    Solar System **89**

Investigation 5: Phases of the Moon
Moonrise/Sunrise Data **90**
Phases of the Moon Sequence Puzzle **91**

Investigation 6: Craters
Archimedes **92**
Aristillus **92**
Lunar Alps **93**
Copernicus **93**
Sea of Serenity **94**
Posidonius **94**
Stöfler **95**
Tycho and Clavius **95**
Barringer Crater, Arizona **96**
Gosses Bluff, Australia **97**
Manicouagan Crater, Canada **98**

Investigation 8: The Solar System
Landforms of the United States **99**
Earth Landforms, Satellite Images **100**
Earth Landforms, Descriptions **104**
Planet Landforms, Images **108**
Planet Landforms, Descriptions **114**

Investigation 9: Space Exploration
Space Missions **120**

Investigation 10: Orbits and New Worlds
Exoplanet Transit Graphs **132**

References
Science Safety Rules **133**
Glossary **134**
Index **136**

White House

White House Neighborhood

White House Community

Washington, DC, Area

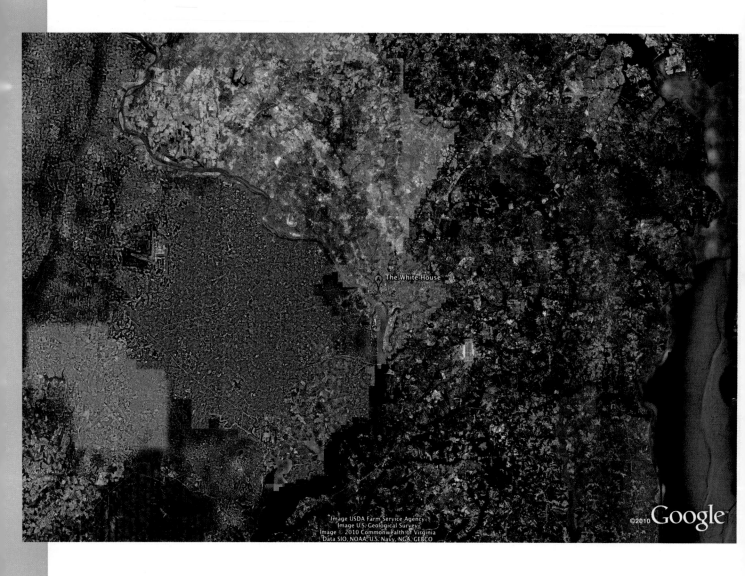

Northeast Region

Earth

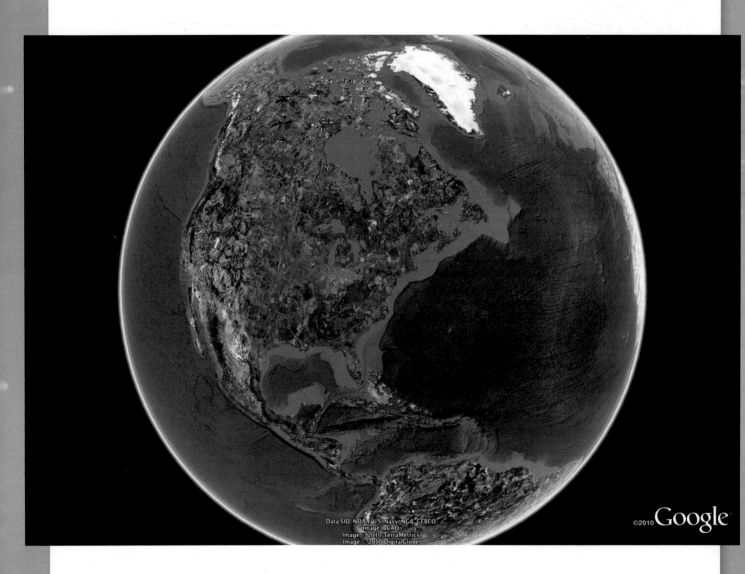

Worldwide Sunrise/Sunset Data

June 21

City and country	Latitude	Sunrise	Sunset	Length of day
Barrow, AK, USA	71° N	None	None	24:00
Stockholm, Sweden	59° N	2:48	8:59	18:11
Sendai, Japan	38° N	4:18	7:06	14:48
Alexandria, VA, USA	38° N	5:44	8:37	14:53
New Delhi, India	28° N	5:22	7:18	13:56
Quito, Ecuador	0°	6:22	6:22	12:00
Nairobi, Kenya	1° S	6:34	6:34	12:00
Auckland, New Zealand	37° S	7:48	4:55	9:07
Punta Arenas, Chile	53° S	8:00	3:32	7:32

December 21

City and country	Latitude	Sunrise	Sunset	Length of day
Barrow, AK, USA	71° N	None	None	0:00
Stockholm, Sweden	59° N	9:53	4:03	6:10
Sendai, Japan	38° N	7:00	4:32	9:32
Alexandria, VA, USA	38° N	7:23	4:50	9:27
New Delhi, India	28° N	7:03	5:31	10:28
Quito, Ecuador	0°	6:14	6:22	12:08
Nairobi, Kenya	1° S	6:24	6:35	12:11
Auckland, New Zealand	37° S	4:56	7:37	14:41
Punta Arenas, Chile	53° S	3:50	8:46	16:56

Think Questions

1. Which locations have the most hours of daylight on June 21? The fewest hours of daylight?

2. Which locations have the most hours of daylight on December 21? The fewest hours of daylight?

3. Alpena, Michigan, is located 45° north of the equator. How much daylight do you estimate it has on June 21? On December 21?

4. Boulder, Colorado, is at latitude 40° north. Wellington, New Zealand, is at latitude 41° south. Which city has the most daylight on June 21?

Investigation 3: *Seasons*

Full Moon

Moon Image

Investigation 4: *Moon Study*

Earth/Moon Comparison

Sun, Planets, and Satellites in the Solar System

Name	Name of planet or satellite	**Radius**	Radius of planet or satellite (in km)	
Orbits	The body it orbits	**Density**	Grams per cubic centimeter	
Distance	Distance to the body it orbits (in 1,000 km)	**Orbital period**	Time to complete orbit (in Earth days)	
		Rotational period	Time to complete rotation (in Earth days)	

Name	Orbits	Distance (1,000 km)	Radius (km)	Density (g/cm^3)	Orbital period	Rotational period
Sun	—	—	695,000	1.41	—	25–36
Mercury	Sun	57,910	2,440	5.42	88.0	59.0
Venus	Sun	108,200	6,052	5.25	225.0	–243.0
Earth	Sun	149,600	6,378	5.515	365.26	1.0
Moon	Earth	384	1,737	3.34	29.5	29.5
Mars	Sun	227,940	3,390	3.94	687.0	1.0
Phobos	Mars	9	14 × 11*	2.0	0.3	0.3
Deimos	Mars	23	8 × 6*	1.7	1.3	1.3
Jupiter	Sun	778,330	71,492	1.33	4,333.0	0.4
Io	Jupiter	422	1,815	3.55	1.8	1.8
Europa	Jupiter	671	1,561	3.01	3.6	3.6
Ganymede	Jupiter	1,070	2,631	1.94	7.0	7.0
Callisto	Jupiter	1,883	2,410	1.86	16.7	16.7
Saturn	Sun	1,429,400	60,268	0.69	10,760.0	0.4
Epimetheus	Saturn	151	72 × 54*	0.7	0.7	0.7
Janus	Saturn	151	98 × 96*	0.67	0.7	0.7
Mimas	Saturn	186	196	1.17	1.0	1.0
Enceladus	Saturn	238	250	1.24	1.4	1.4
Tethys	Saturn	295	533	1.21	1.9	1.9
Dione	Saturn	377	560	1.43	2.7	2.7
Rhea	Saturn	527	765	1.33	4.5	4.5
Titan	Saturn	1,222	2,575	1.88	16.0	16.0
Iapetus	Saturn	3,561	730	1.21	79.0	79.0
Uranus	Sun	2,870,990	25,559	1.29	30,685.0	–0.7
Miranda	Uranus	130	236	1.15	1.4	1.4
Ariel	Uranus	191	579	1.56	2.5	2.5
Umbriel	Uranus	266	585	1.52	4.0	4.0
Titania	Uranus	436	789	1.7	8.7	8.7
Oberon	Uranus	583	775	1.64	13.5	13.4
Neptune	Sun	4,501,200	24,622	1.64	60,190.0	0.7
Triton	Neptune	355	1,350	2.07	–5.9	–5.9

* This measurement is length × width because this satellite is not a sphere.

Moonrise/Sunrise Data

Here are the moonrise and sunrise times for January 2011 for Berkeley, California.

Date	Moonrise	Sunrise	Phase
January 1	5:03 a.m.	7:25 a.m.	
January 2	6:02 a.m.	7:25 a.m.	
January 3	6:53 a.m.	7:25 a.m.	
January 4	7:37 a.m.	7:25 a.m.	New Moon
January 5	8:15 a.m.	7:25 a.m.	
January 6	8:47 a.m.	7:25 a.m.	
January 7	9:15 a.m.	7:25 a.m.	
January 8	9:41 a.m.	7:25 a.m.	
January 9	10:05 a.m.	7:25 a.m.	
January 10	10:30 a.m.	7:25 a.m.	
January 11	10:56 a.m.	7:25 a.m.	
January 12	11:24 a.m.	7:24 a.m.	First quarter
January 13	11:56 a.m.	7:24 a.m.	
January 14	12:34 p.m.	7:24 a.m.	
January 15	1:18 p.m.	7:24 a.m.	
January 16	2:11 p.m.	7:23 a.m.	
January 17	3:12 p.m.	7:23 a.m.	
January 18	4:20 p.m.	7:23 a.m.	
January 19	5:32 p.m.	7:22 a.m.	Full Moon
January 20	6:46 p.m.	7:22 a.m.	
January 21	7:59 p.m.	7:21 a.m.	
January 22	9:11 p.m.	7:21 a.m.	
January 23	10:23 p.m.	7:20 a.m.	
January 24	11:33 p.m.	7:19 a.m.	
January 25		7:19 a.m.	
January 26	12:43 a.m.	7:18 a.m.	Third quarter
January 27	1:52 a.m.	7:17 a.m.	
January 28	2:57 a.m.	7:17 a.m.	
January 29	3:56 a.m.	7:16 a.m.	
January 30	4:49 a.m.	7:15 a.m.	
January 31	5:35 a.m.	7:14 a.m.	

Phases of the Moon Sequence Puzzle

Starting with the full Moon, put the images in order to show the sequence of the phases of the Moon.

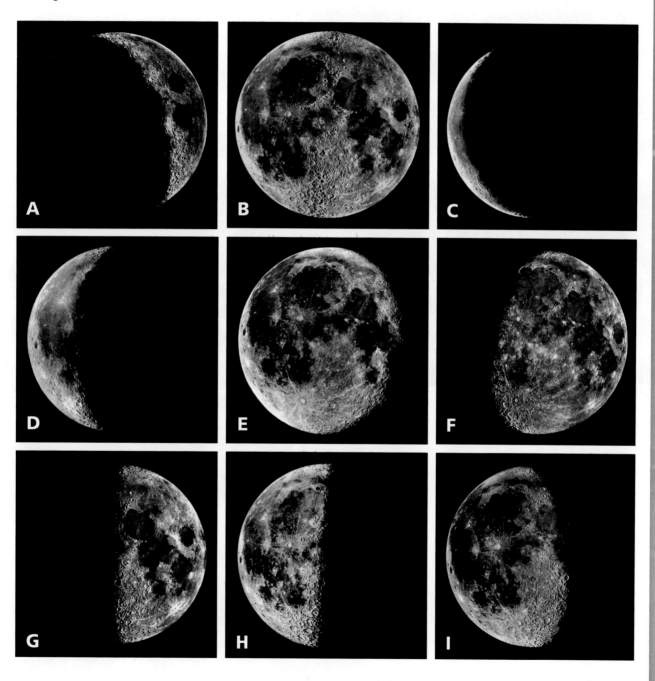

Archimedes

Aristillus

Lunar Alps

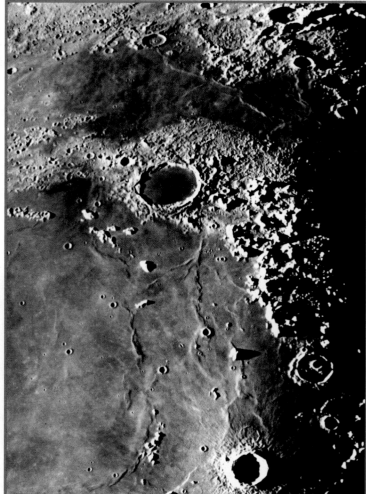

Copernicus

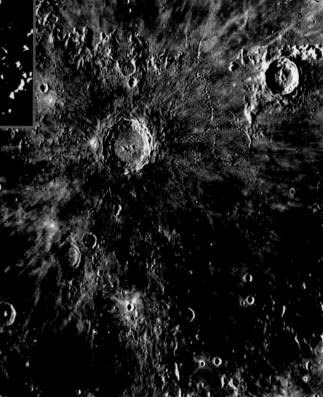

Sea of Serenity

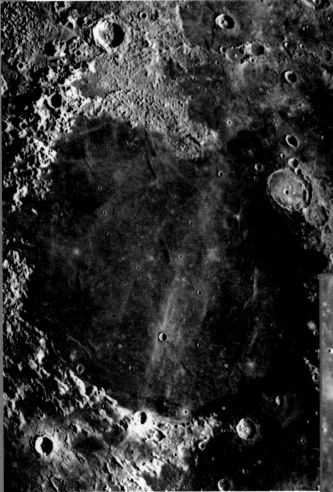

Posidonius

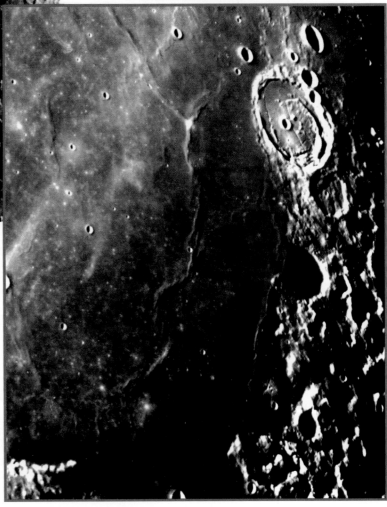

Stöfler

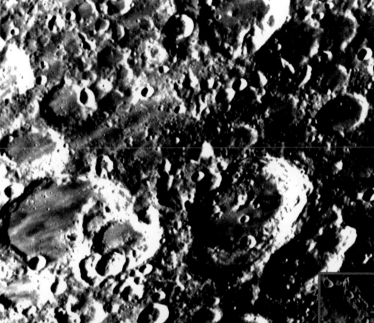

Tycho and Clavius

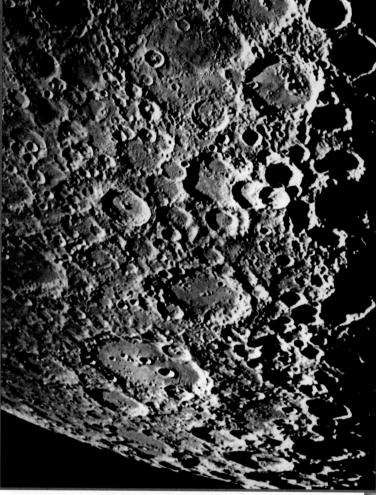

Investigation 6: Craters

Barringer Crater, Arizona

Gosses Bluff, Australia

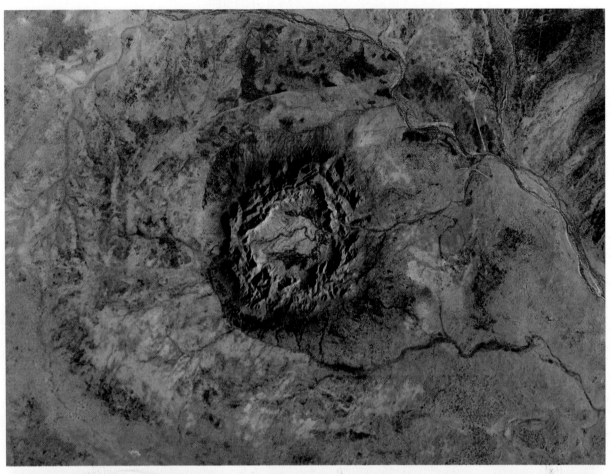

Investigation 6: *Craters*

Manicouagan Crater, Canada

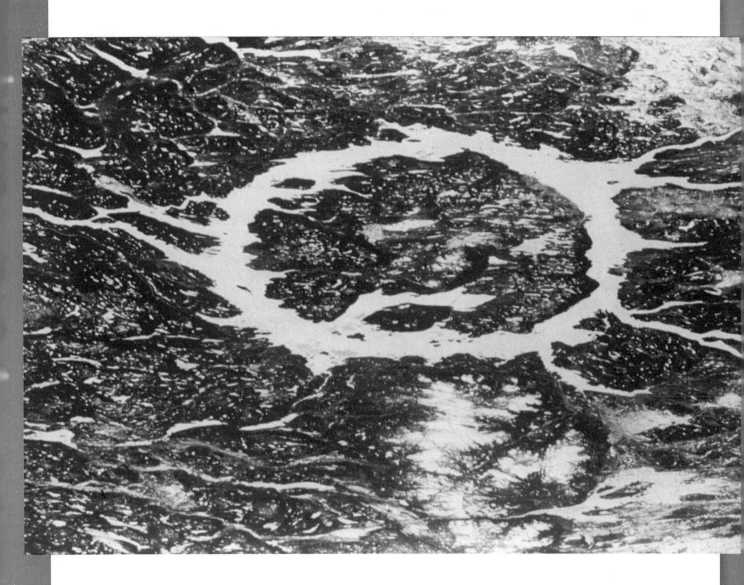

Landforms of the United States

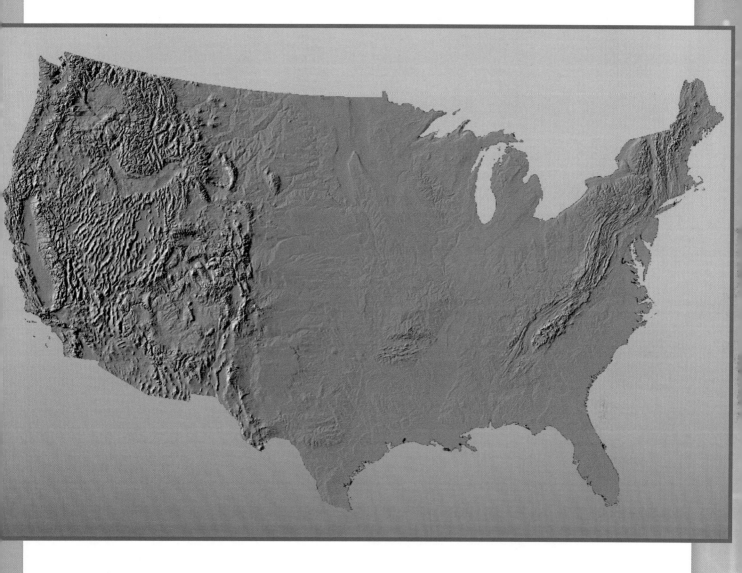

Investigation 8: *The Solar System* **99**

Earth Landforms, Satellite Images

A. Brazil

B. Brazil

C. Chile

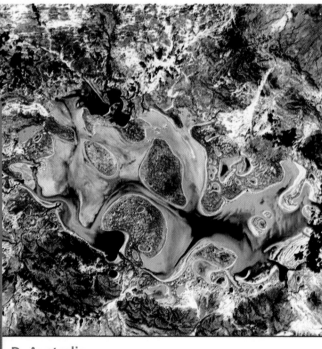

D. Australia

E. Louisiana, USA

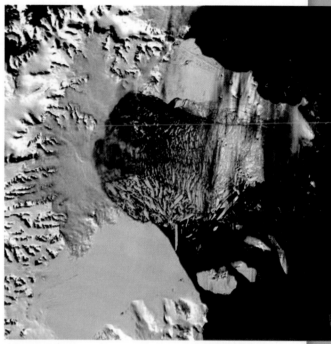
F. Antarctica

G. Washington, USA

H. Egypt

Investigation 8: **The Solar System** **101**

I. Argentina

J. Australia

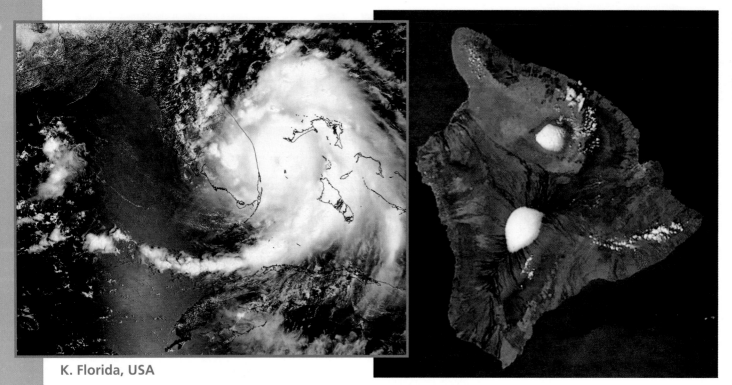

K. Florida, USA

L. Hawaii, USA

M. Venezuela

N. California, USA

O. USA and Canada

P. Tibet

Investigation 8: *The Solar System* **103**

Earth Landforms, Descriptions

A. Brazil

Twists and turns in this riverbed create islands that might be covered with the next flood. The riverbed for a meandering river is constantly shifting.

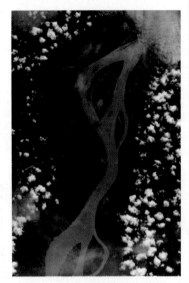

This image of a river in Brazil was taken from the International Space Station from an altitude of 383 km on March 11, 2002.

B. Brazil

Close to the city of Manaus, Brazil, the Rio Solimões and Rio Negro converge to form the Amazon River. Manaus is the gray patch to the right of the image's center.

The water of the Rio Solimões is pale and murky because it carries glacial silt and sand from its origin in the Peruvian Andes. The dark color of the Rio Negro is a feature of clear waters that carry little sediment and come from areas of bedrock. East of Manaus, the pale and dark waters flow side-by-side as distinct flows before they eventually merge and mix. Northwest of Manaus, on the Rio Negro, is the Anavilhanas archipelago, the largest group of freshwater islands in the world.

C. Chile

Chiliques Volcano in Chile, thought to be dormant for 10,000 years, is now categorized as active after NASA satellite data revealed hot spots in the crater. Chiliques is a stratovolcano

with a circular summit crater at an elevation of 5,778 m. The crater is 500 m in diameter. This mountain is one of the most important high-altitude ceremonial centers of the Incas. People rarely visit it because it is so difficult to get to. As you climb to the summit along Inca trails, you pass numerous ruins. At the summit is a series of constructions used for rituals. A beautiful lagoon in the crater is almost always frozen.

D. Australia

Lake Carnegie in Western Australia fills with water only during periods of significant rainfall. In dry years, it is a muddy marsh.

This is a false-color composite image made using shortwave infrared, infrared, and red wavelengths.

E. Louisiana, USA

This image includes the coastal areas of Louisiana, Mississippi, Alabama, and part of the Florida Panhandle on the Gulf of Mexico. It covers an area 345 km by 315 km. New Orleans is visible at the southern edge of Lake Pontchartrain, along the left-hand side.

The Mississippi River delta, in the center, is called a bird's-foot delta because of its distinctive shape. Variations in ocean color show suspended sediment from the river and coastal areas as water flows into the Gulf of Mexico. Large amounts of sediment from the land have been deposited in shallow coastal waters. These delta environments form channels and coastal wetlands that provide important habitats for waterfowl and fisheries. The city of New Orleans is prone to flooding, with about 45 percent of the city situated at or below sea level. Levees and wetlands help buffer the city from storm surges.

F. Antarctica

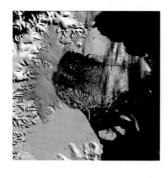

The Larsen Ice Shelf is a large, floating ice mass on the eastern side of the Antarctic peninsula. Antarctic ice shelves are thick plates of floating ice fed by Antarctic continental glaciers. In a 35-day period beginning on January 31, 2002, about 3,250 square km of the shelf disintegrated. This event sent thousands of icebergs adrift into the Weddell Sea. Over the next 5 years, the shelf lost a total of 5,700 square km, about 60 percent of its previous size.

G. Washington, USA

Mount St. Helens in western Washington State is a stratovolcano. After lying dormant for 10,000 years, its last major eruption occurred on May 18, 1980. It is the most active volcano in the Cascade Range. This image was captured on March 15, 2005, one week after an ash and steam eruption. A new lava dome in the southeast part of the crater is clearly visible. It is highlighted by red areas where the Advanced Spaceborne Thermal Emission and Reflection Radiometer's (ASTER) infrared channels detected hot spots from incandescent lava. The new lava dome is 155 m higher than the old lava dome and still growing.

H. Egypt

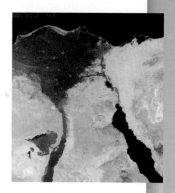

The northern portion of the Nile Delta shows the triangle shape of the Greek letter delta, Δ. The Nile is the longest river in the world. It extends about 6,825 km from its headwaters in the highlands of eastern Africa to the Mediterranean Sea.

At the bottom of the fertile Nile Delta is Cairo, the capital of Egypt. To the west are the Great Pyramids of Giza. North of here the Nile branches into two distributaries, the Rosetta to the west and the Damietta to the east.

Also visible in this image is the Suez Canal, a shipping waterway connecting Port Said on the Mediterranean Sea with the Gulf of Suez. The gulf is an arm of the Red Sea and is located on the right-hand side of the image.

I. Argentina

This region of the Andes mountains is south of San Martín de Los Andes, Argentina. It shows the steep-sided valleys and other distinctive landforms carved by Pleistocene glaciers. Elevations here range from about 700 to 2,440 m. Tectonic and volcanic activity is very common in this region. The landforms provide a record of the changes that have occurred over many thousands of years. Large lakes fill the broad mountain valleys, and the spectacular scenery makes this area a popular resort destination for Argentinians.

This false-color image was acquired by the first Shuttle Radar Topography Mission (SRTM).

What do you think the colors in this image represent?

J. Australia

The Great Barrier Reef extends for 2,000 km along the northeastern coast of Australia. It is not a single reef, but a vast maze of reefs, passages, and coral cays (islands that are part of the reef). The large island off the most northerly part of the coast is Whitsunday Island. Smaller islands and reefs extend southeast, parallel to the coast.

This true-color image shows part of the southern portion of the reef approximately 380 km wide. The reef is next to the central Queensland coast, which is the land to the west of the reef. The reef is clearly visible up to approximately 200 km from the coast.

K. Florida, USA

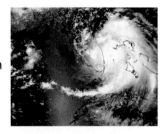

Hurricane Katrina was a relatively weak Category 1 hurricane as it approached Florida on August 24, 2005. After it crossed the Florida peninsula, the warm Gulf of Mexico waters caused it to grow in 9 hours to a Category 5 hurricane. On August 29, this powerful hurricane struck the Louisiana coast and caused an estimated $81 billion in damages and killed at least 1,836 people.

An outline of Florida, the Bahamas, and the Gulf coastline have been added to this image.

L. Hawaii, USA

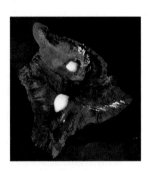

On January 14, 2005, white snow covered the summits of the high peaks of Hawaii's Mauna Loa (south) and Mauna Kea (north) volcanoes. The third volcano that makes up the island is Kilauea. Kilauea is currently active, so the lava flows give off heat, which can be detected by certain satellites (area outlined in red).

Dark streaks of cooled and hardened lava reach down the brown slopes of the volcanoes toward the ring of lush forests of the lower slopes. The dots of gray around the notch in the middle of the northeastern coastline is Hawaii's largest city, Hilo.

Of the three volcanoes on Hawaii, Kilauea is the most active, having started its most recent eruption in 1983. It hasn't stopped since. The majority of the surface of the volcano is covered by lava flows less than 1,000 years old, which reveal how active the volcano has been.

Answer for I: The colors represent the different elevations. The colors from lowest to highest elevation are green, yellow, red, magenta, and white.

M. Venezuela

Duckweed plants form green swirls on the surface of Lake Maracaibo in northern Venezuela. The lake is usually too salty to support duckweed. But in 2004, unusually heavy rain brought additional fresh water to the lake. The fresh water stirred the nutrient layers below and allowed the nutrients to float to the surface with the less dense freshwater layer on top. For a brief time, the duckweed used the nutrients to grow, and it doubled in area every day. Then, as the lake began to settle back into its normal layers with nutrient-rich waters on the salty bottom, the duckweed's growth slowed and eventually stopped.

N. California, USA

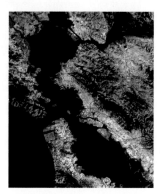

This image of the San Francisco Bay region shows the thermal (heat) radiation differences in urban building and road materials. The reds show more heat emissions, and the greens show less heat emissions. The water of the San Francisco Bay and Pacific Ocean (lower left) appear black.

O. USA and Canada

The Great Lakes are the largest collection of freshwater lakes on Earth. They formed about 10,000 years ago at the end of the last ice age. The lakes were carved out by the southern movement of large glaciers.

The Great Lakes lie on the border between Canada to the north and the United States to the south.

In this mostly cloud-free image, areas with sediments flowing into the lakes are clearly seen. The light color shows sediment in the southern portion of Lake Huron, Lake St. Clair, much of Lake Erie, and the outflow of the Niagara River into Lake Ontario.

P. Tibet

The Kunlun Fault is one of the gigantic strike-slip fault systems that bound the north side of Tibet and the Himalayas. Motion along the 1,500 km length of the Kunlun has occurred for the last 40,000 years at a rate of 1.1 cm per year, creating an offset of more than 400 m.

In this image, two faults are seen from east to west. The northern fault is marked by lines of vegetation, which appear red in the image. The southern, younger fault cuts through the alluvial fans. These alluvial fans form when fast-flowing water from the mountains spreads out and drops sediments as it reaches the valley floor.

Planet Landforms, Images

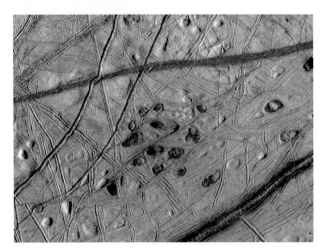
1. Jupiter's moon Europa

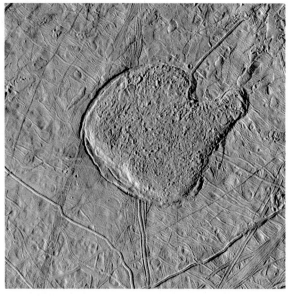
2. Jupiter's moon Europa

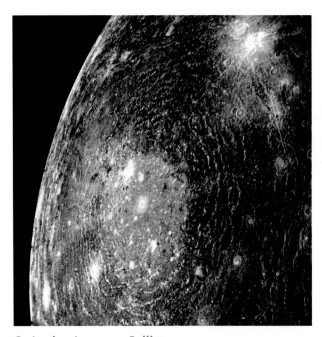
3. Jupiter's moon Callisto

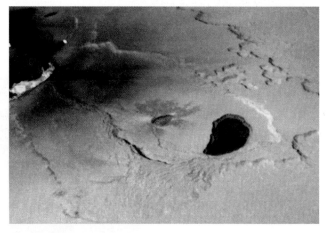
4. Jupiter's moon Io

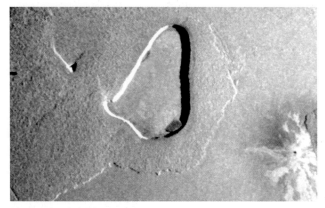
5. Jupiter's moon Io

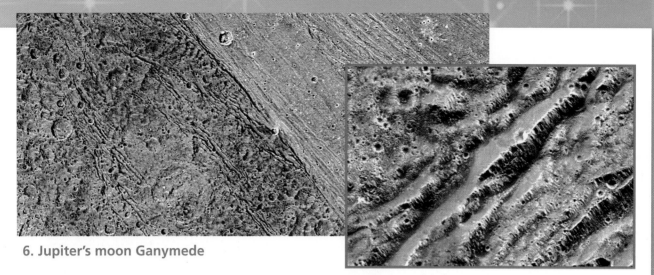
6. Jupiter's moon Ganymede

7. Ganymede close-up

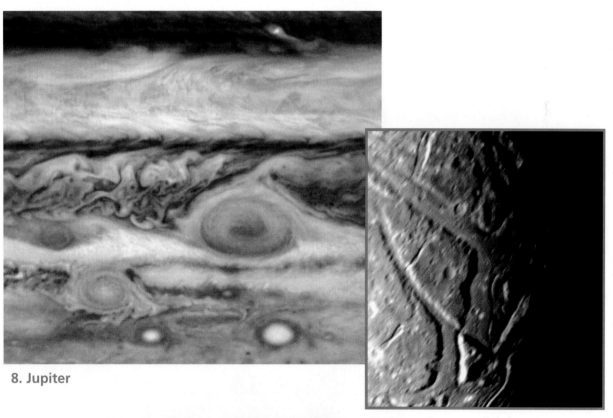
8. Jupiter

9. Uranus's moon Ariel

10. Uranus's moon Miranda

Investigation 8: The Solar System **109**

11. Saturn

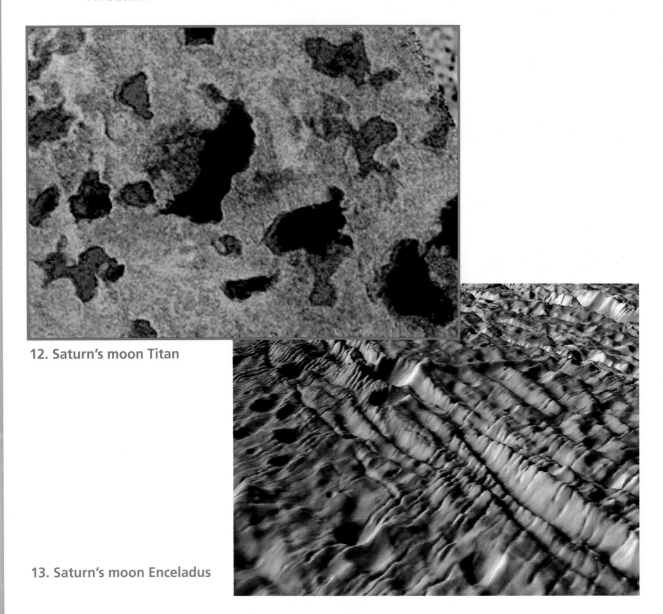

12. Saturn's moon Titan

13. Saturn's moon Enceladus

110

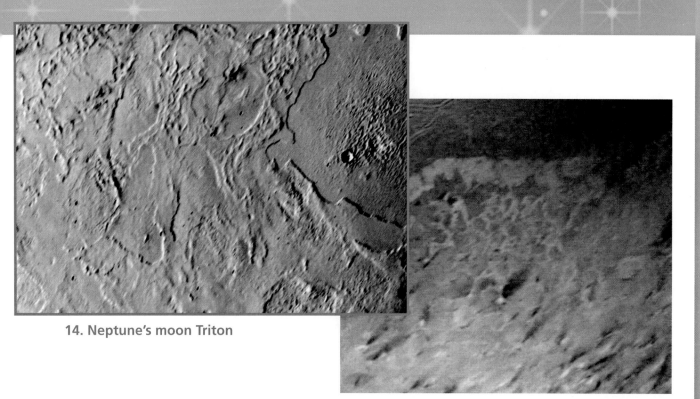

14. Neptune's moon Triton

15. Triton close-up

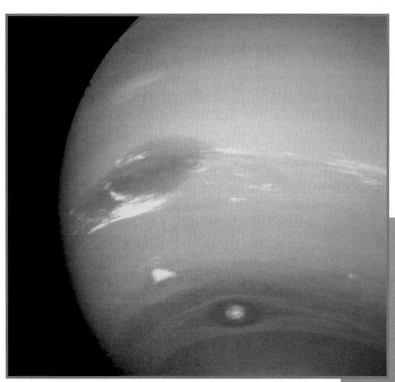

16. Neptune

17. Neptune

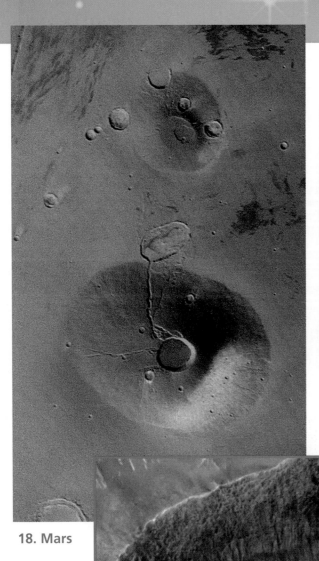
18. Mars

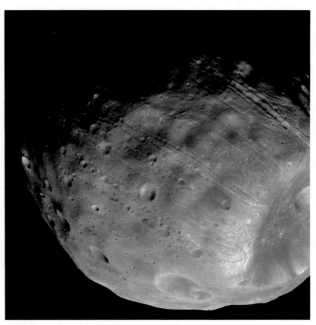
19. Mars's moon Phobos

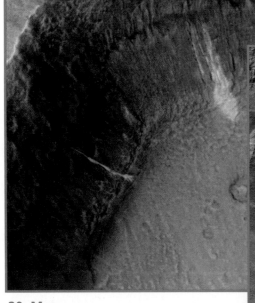
20. Mars

21. Mars 2 km

22. Mercury

23. Mercury close-up

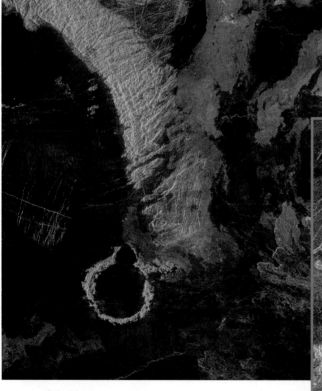
24. Venus

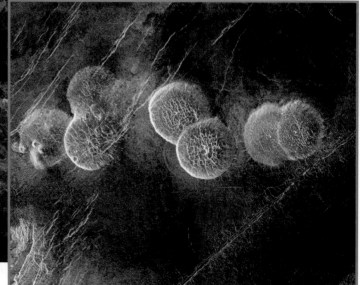
25. Venus close-up

*Investigation 8: **The Solar System**

Planet Landforms, Descriptions

1. Jupiter's Moon Europa

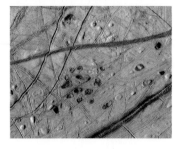

Reddish spots and shallow pits dot the icy surface of Europa. These pits might be the result of warm ice bubbling up through the colder icy shell around Europa. Data suggest that there is a deep liquid-water ocean beneath the ice layer. The red color may hold clues to the composition of the ocean and to whether it could support life.

The fractures and cracks that crisscross the surface also point to a liquid layer below. Water movement cracks the surface, and gaps fill with water and quickly refreeze.

2. Jupiter's Moon Europa

This *Galileo* spacecraft view of Jupiter's icy moon, Europa, shows a region shaped like a mitten or catcher's mitt. It has a texture that is seen in numerous locations across Europa's surface.

The material in the "catcher's mitt" has the appearance of frozen slush (water) and seems to bulge upward from the touching surface. Scientists on the Galileo imaging team are exploring various hypotheses for the processes that formed these surface textures. These processes include convection (vertical movement between areas that differ in density due to heating), upwelling of viscous icy "lava," or liquid water melting through from a subsurface ocean.

3. Jupiter's Moon Callisto

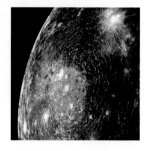

This multi-ring basin on Jupiter's icy moon Callisto is similar to the large circular impact basins that dominate the surface of Earth's Moon and the planet Mercury. The inner parts of these basins are generally surrounded by ejecta and several mountainous ring structures that are thought to have formed during the impact. The ring structures on the Moon and Mercury have been likened to ripples produced on a pond by a rock striking the water. The great number of rings observed around this basin on Callisto is consistent with a low-density layer below the surface. Examination of close-up images suggest that there could be an ocean of liquid water beneath this icy surface.

4. Jupiter's Moon Io

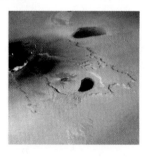

An active volcanic eruption on Jupiter's moon Io was captured in this image taken on February 22, 2000, by NASA's *Galileo* spacecraft. White and orange areas on the left side of the photo show newly erupted hot lava. The two small bright spots are sites where molten rock is exposed to the surface at the edge of lava flows. The larger orange and yellow ribbon is a cooling lava flow that is more than 60 km long. The orange, yellow, and white areas show temperature variations, orange being the coolest and white the hottest material. Dark deposits surrounding the active lava flows were not there during a November 1999 flyby of Io.

The lava on Io is much hotter than lava produced on Earth. Io is considered to be the most active volcanic body in the solar system because of the amount of heat that its volcanoes produce.

5. Jupiter's Moon Io

This image taken on October 16, 2001, shows different volcano types on Io. The center shows a large volcanic depression, or patera, almost 100 km long. It may have formed after eruptions of lava emptied a magma chamber, leaving a space into which the crust collapsed. Evidence of lava flows associated with this patera, however, is scarce. Either the flows were buried, or they never erupted above the surface.

To the right is a shield volcano, similar to volcanoes in Hawaii. Lava thick enough to pile up into shields is rare on Io. Lava on Io usually runs out in long, thin flows. These lava flows could be made of sulfur.

6–7. Jupiter's Moon Ganymede

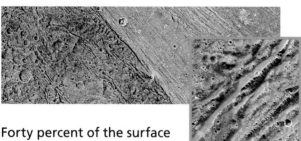

Forty percent of the surface of Ganymede is covered by highly cratered dark regions. The remaining 60 percent is covered by a light grooved terrain, which forms intricate patterns. The dark regions are old and rough and believed to be the original crust of Ganymede. Lighter regions are young and smooth (unlike Earth's Moon).

Since Ganymede has a low density, it was originally estimated that it is half ice with a rocky core extending to half of its radius. However, the *Galileo* spacecraft found a magnetic field around Ganymede, which strongly indicates that it has a metallic core about 40–1,280 km below the surface. The mantle is composed of ice and silicates and a crust that is probably a thick layer of ice.

8. Jupiter

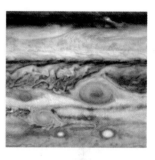

Jupiter's most outstanding surface feature is the Great Red Spot, a swirling mass of gas that looks like a hurricane. The widest diameter of the spot is about three times that of Earth. The color of the spot usually varies from brick-red to slightly brown. Rarely, the spot fades entirely. Its color may be due to small amounts of sulfur and phosphorus in ammonia crystals.

The atmosphere of Jupiter is composed of about 86 percent hydrogen, 14 percent helium, and tiny amounts of methane, ammonia, phosphine, water, acetylene, ethane, germanium, and carbon monoxide. Scientists have calculated these amounts from measurements taken with telescopes and other instruments on Earth and aboard spacecraft. The highest white clouds are made of frozen ammonia crystals. Darker, lower clouds of other chemicals occur in the belts. Astronomers had expected to detect water clouds about 70 km below the ammonia clouds. However, none have been discovered at any level.

9. Uranus's Moon Ariel

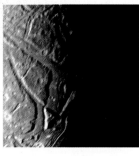

This high-resolution view of Uranus's icy moon Ariel shows complex crisscrossing valleys with impact craters on top. *Voyager 2* obtained this narrow-angle view from a distance of 130,000 km and with a resolution of about 2.4 km. Particularly striking to Voyager scientists is that the faults crossing the valleys are not visible in the bottoms of the valleys they cross. Apparently these valleys were filled with deposits sometime after they were formed, leaving them flat and smooth. Trenches later formed, probably by some flow process.

10. Uranus's Moon Miranda

Miranda, innermost of Uranus's large, icy moons, is seen at close range in this image, taken January 24, 1986, as part of a high-resolution mosaic sequence. *Voyager 2* was some 36,000 km away from Miranda. This clear-filter, narrow-angle view shows an area about 250 km across, at a resolution of about 800 m. Two distinct terrain types are visible: a rugged, higher-elevation terrain (right) and a lower, streaked terrain. Numerous craters on the higher terrain indicate that it is older than the lower terrain. Several slopes, probably faults, cut the different terrains. The impact crater in the lower part of this image is about 25 km across.

11. Saturn

As a gas giant planet, Saturn is mostly atmosphere. The atmosphere is hydrogen and helium but also contains traces of ammonia, phosphine, methane, and other compounds.

Saturn has powerful lightning storms that are 10,000 times stronger than on Earth. They occur in huge, deep columns nearly as large as Earth. The storms occasionally burst through to the planet's visible cloud tops.

In this image taken by the *Cassini* spacecraft, large swirling storms edge their way along the boundary between cloud bands that flow east/west.

12. Saturn's Moon Titan

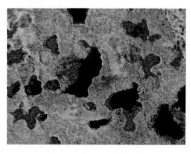

This false-color image of Titan shows bodies of liquid near the north pole. These have features that are commonly associated with lakes on Earth, such as islands, bays, inlets, and channels. The lakes are most likely liquid methane and ethane.

The lakes on Titan are widespread, and at least one lake is larger than Lake Superior on the US/Canadian border. It covers an area of about 100,000 square km. Analysis of the data indicates that the bodies of liquid may be tens of meters in depth.

13. Saturn's Moon Enceladus

The surface of Enceladus is one of the brightest objects in the solar system. It is covered by ice that reflects sunlight like freshly fallen snow.

There are at least five different types of terrains seen on Enceladus including fissures, plains, and crustal deformation that indicates the interior may be liquid.

14–15. Neptune's Moon Triton

A fresh impact crater can be seen in the large, smooth area on the right side of this image of Triton. The low cliffs to the left may have been caused by melting surface materials or fluids that flowed in the past.

The small white spots in this close-up of Triton appear to be frost around volcanic vents. This false-color image of Triton was taken when *Voyager 2* was about 190,000 km from Triton's surface.

16. Neptune

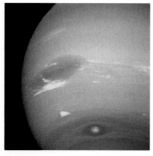

Neptune's atmosphere is similar to the other gas giant planets in the solar system. It consists mainly of hydrogen and helium with trace amounts of methane, water, and ammonia. But Neptune has a larger portion of methane ice in its upper atmosphere, which gives it a distinctive blue color.

Neptune has a long-lasting storm, called the Great Dark Spot, similar to the Great Red Spot on Jupiter. The fastest winds in the solar system have been clocked on Neptune at 2,400 km per hour.

The small white spot below the Great Dark Spot is a bright feature the Voyager scientists nicknamed Scooter.

17. Neptune

Voyager 2 captured this image of clouds in the outer atmosphere of Neptune. The cloud streaks range from 30 to 50 km wide.

The white, upper clouds consist mainly of frozen methane. The darker, lower clouds are likely made of hydrogen sulfide.

Cloud heights appear to be approximately 50 km above the surface.

Investigation 8: The Solar System

18. Mars

This image shows two Martian volcanoes: Ceraunius Tholus (lower) and Uranius Tholus (upper). Impact craters on the slopes of these volcanoes indicate they are old and inactive.

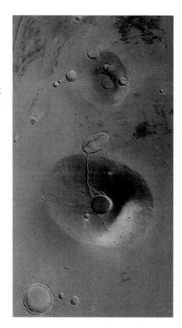

The crater at the summit of Ceraunius Tholus is about 25 km across. Remains of an ancient flow can be seen on the northern slope.

19. Mars's Moon Phobos

The most striking feature on Phobos, Mars's larger moon, is Stickney crater. It appears in the lower right part of the image. The crater is about 9 km across.

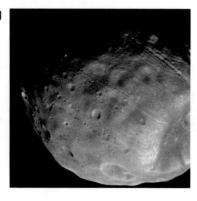

Light-colored streaks on the inside of Stickney are landslides. Troughs and crater chains are visible outside the crater. They are probably the result of material ejected from impacts on Mars that flew off and hit Phobos.

20. Mars

Has liquid water flowed on Mars in recent times? This image taken in June 2000 looks like a water flow on the side of this crater. Since that time, tens of thousands of slopes have been imaged by all the Mars-orbiting spacecraft to see if anything changed. Mars scientists have identified changes.

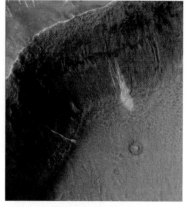

Does this image prove that water recently flowed on Mars? No, but it provides the first evidence that this might have occurred. The surface on Mars is drier than the most arid deserts on Earth. But liquid water from beneath the Martian surface might still periodically come to the surface and flow across the red planet.

21. Mars

This image shows a fossil delta in Eberswalde crater on Mars. The delta shows the first clear evidence that liquid water flowed on the surface of Mars. Unlike the recently observed changes, this delta is old. The sediments were deposited in water, covered by more sediments, and turned to rock. Wind erosion has exposed this ancient delta.

22. Mercury

This area on the floor of the Caloris basin (Caloris Planitia) was nicknamed the Spider by the team that studied it. What looks like a set of troughs radiating from the center has been interpreted as evidence that the basin floor broke apart during the impact that formed the crater.

23. Mercury

Details of Mercury's surface features are visible in this image taken during the MESSENGER Mission's closest approach to the planet. In addition to craters as small as 400 m, one of the highest cliffs found on Mercury sweeps from the top center to the left side of the image. The cliffs must have been formed by tremendous forces in Mercury's crust.

24. Venus

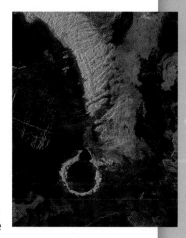

The oldest rocks in this false-color radar image of Leda Planitia (plains) of Venus are the bright, highly fractured highlands in the upper left. The darker plains that surround the highlands formed when volcanic lava filled the low-lying areas. Plains like this cover as much as 80 percent of the surface of Venus. The most recent lava flows are the bright areas in the upper right.

The circular structure in the lower left is probably an impact crater. Venus is struck by meteors at about the same rate as Earth.

25. Venus

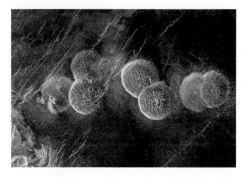

The seven circular domelike hills dominate the eastern edge of Alpha Regio on Venus, as shown in the radar image. The domes average 25 km in diameter with maximum heights of 750 m. They probably formed when thick lava erupted through the flat plains and flowed outward. These domes might be similar to volcanic domes on Earth. Another interpretation suggests that the lava pushed up beneath the surface but never erupted onto the plains.

Investigation 8: The Solar System **119**

Space Missions

Humans have always looked to the sky, wondered what was out there, and dreamed of visiting worlds beyond our own. However, sending humans into the universe is not a viable option. Other than a few trips to the Moon, we have been bound within the upper reaches of the atmosphere.

For over 45 years, NASA has been sending robotic probes into space, to go where no human has gone before. These robotic spacecraft act like extensions of humans, collecting information far out in the solar system (but not beyond), and sending it back to Earth. Space probes have parts that mimic and extend human senses and abilities. Even if human astronauts can't visit other planets yet, robotic probes can provide information that is the next best thing to being there. The table compares human body parts with their spacecraft counterparts. But it is not completely filled in.

Can you think of the missing spacecraft parts that function like our body parts?

Navigation and Orientation

Imagine a spacecraft that has traveled millions of kilometers through space for a first-ever flyby of a planet. But its cameras are pointed in the wrong direction as it speeds past! People on Earth would be very upset.

To help prevent such a disaster, spacecraft sight on the Sun and a bright star, such as Canopus, to help navigate and maintain orientation. If the probe is off course or pointing in the wrong direction, small rocket engines on the spacecraft can make adjustments in the flight path and orientation.

Probes communicate with Earth using radio signals. The radio signal is weak if the probe is millions of kilometers from Earth. So NASA ground controllers use three huge "ears," steerable radio antennas (Deep Space Network) located in California, Australia, and Africa. The network listens for the signal, which is then sent to the probe's crew at a NASA mission-control center, such as Jet Propulsion Laboratory (JPL).

Human body	Spacecraft counterpart
Body/torso	The housing holds the spacecraft components and attaches to other devices. Also known as the bus.
Neck	The scan platform turns so that the instruments can point in the desired direction without reorienting the whole spacecraft.
Brain	
Nerves	
Skin	Blankets protect against meteorites and help control temperature. The spacecraft can't sweat, so it uses radiators to get rid of excess heat.
Legs	Rocket motors change a spacecraft's orientation and course.
Blood vessels	
Feet	
Arms	
Sense organs (eyes, ears, nose, taste buds, touch sensors)	
Voice	

Asteroid Missions

Near Earth Asteroid Rendezvous–Shoemaker

Acronym: NEAR Shoemaker

Destination: 433 Eros

Goals: Collect data on the properties, composition, surface features, interior, and magnetic field of Eros. Study surface properties, interactions with solar wind, possible currents of dust or gas, and asteroid spin. Look for clues about the formation of Earth and other planets.

Launch: February 17, 1996

Arrival: Orbit February 14, 2000; touchdown February 12, 2001

End of Mission: Last message February 28, 2001

Hayabusa

Destination: 25143 Itokawa

Goals: Collect a surface sample of material from the small (550 m × 180 m) asteroid 25143 Itokawa and return the sample to Earth. Record detailed studies of the asteroid's shape, spin, topography, color, composition, density, brightness, interior, and history.

Launch: May 9, 2003

Arrival: September 12, 2005

End of Mission: June 13, 2010

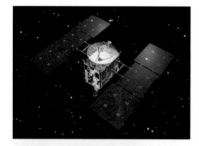

Dawn

Destination: Vesta and Ceres

Goals: Learn the conditions and processes of the early solar system by detailed investigation of asteroids Vesta and Ceres, two of the largest protoplanets not broken up since their formation. Determine how protoplanet size and water content are involved in the evolution of planets.

Launch: September 27, 2007

Arrival: Vesta orbit 2011–2012; Ceres orbit 2015

End of Mission: Still operating as of 2012

Comet Missions

Stardust

Destination: Comet Wild 2

Goals: Collect comet dust during a close encounter with Comet Wild 2 (pronounced "Vilt 2"). Collect interstellar dust.

Launch: February 7, 1999

Arrival: January 2, 2004

End of Mission: January 15, 2006

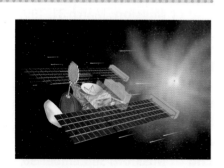

Deep Space 1
Acronym: DS1
Destination: Comet Borrelly
Goals: Test 12 new, high-risk technologies. These include ion propulsion (rocket engine using ionized xenon gas), Autonav (automatic navigation system), Remote Agent (remote intelligent self-repair software), Small Deep-Space Transponder (miniaturized radio system and instruments), Solar Concentrator Array, Beacon Monitor experiment, and others.
Launch: October 24, 1998
Arrival: September 22, 2001
End of Mission: December 18, 2001

Rosetta Orbiter
Destination: Comet 67P/Churyumov-Gerasimenko
Goals: Study the origin of comets. Study the relationship between comets and other interstellar material and what this tells us about the origin of the solar system. Describe the comet nucleus, surface features, and composition.
Launch: March 2, 2004
Arrival: May 2014
End of Mission: Still operating as of 2012

Outer Solar System Missions

Deep Impact
Destination: Comet Tempel 1
Goals: What is the original material in comets? Do comets lose their ice or seal it in and become dormant? What do we know about crater formation?
Launch: January 12, 2005
Arrival: Comet Tempel 1, July 4, 2005
End of Mission: August 2005

Pioneer 10
Destination: Jupiter
Goals: Investigate Jupiter system, the constellation Taurus, and beyond
Launch: March 2, 1972
Arrival: December 3, 1973
End of Mission: March 31, 1997

Pioneer 11
Destination: Jupiter and Saturn
Goals: Investigate Jupiter, Saturn, and the outer solar system. Study energetic particles in the outer solar system (heliosphere).
Launch: April 5, 1973
Arrival: Jupiter, 1974; Saturn, 1979
End of Mission: September 30, 1995

Juno

Destination: Jupiter

Goals: Build on the results of previous missions and provide new information to help determine how, when, and where this gas giant planet formed. Collect data from highly elliptical polar orbits (traveling over the north pole to the south pole and back to the north pole) that skim only 5,000 km above the planet's atmosphere. Improve our understanding of the origin of the solar system.

Launch: August 5, 2011

Arrival: July 2016

End of Mission: Still operating as of 2012 (scheduled to end in 2017)

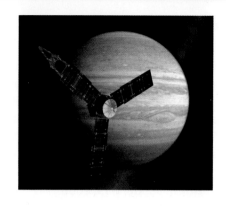

Voyager 1 and 2

Destination: Jupiter, Saturn, Uranus, and Neptune

Goals: Conduct close-up studies of Jupiter and Saturn, Saturn's rings, and the larger moons of both planets. *Voyager 2* was extended to include flybys of Uranus and Neptune. Both *Voyagers 1* and *2* will explore the boundaries of our solar system and travel into interstellar space.

Launch: *Voyager 2*, August 20, 1977; *Voyager 1*, September 5, 1977

Arrival: *Voyager 1*: Jupiter, 1979; Saturn, 1980; outer solar system, 2004; *Voyager 2*: Jupiter, 1979; Saturn, 1981; Uranus, 1986; Neptune, 1989; outer solar system, 2007

End of Mission: Still operating as of 2012

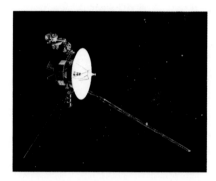

Galileo

Destination: Jupiter

Goals: Collect data on the magnetosphere and study Jupiter and its moons. Study Venus, Earth, the Moon, and two asteroids, Gaspra and Ida, during flybys.

Launch: October 18, 1989

Arrival: Probe released July 13, 1995; probe entered atmosphere December 7, 1995

End of Mission: September 21, 2003

Cassini-Huygens

Destination: Saturn

Goals: Conduct a detailed study of Saturn and its moons, including temperature and composition of Saturn's atmosphere, cloud properties, winds, internal structure, its rings, icy moons, and Titan.

Launch: October 15, 1997

Arrival: June 30, 2004

End of Mission: Still operating as of 2012 (mission extended through 2017)

Investigation 9: Space Exploration

New Horizons

Destination: Pluto

Goals: Study the dwarf icy planets at the edge of the solar system and visit Pluto and Charon for the first time. Map surface composition and characterize the geology of Pluto and Charon. Describe the atmosphere of Pluto and search for an atmosphere around Charon. Map surface temperatures on Pluto and Charon and search for rings and additional moons around Pluto. Conduct similar investigations of one or more Kuiper Belt objects.

Launch: January 19, 2006

Arrival: July 2015

End of Mission: Still operating as of 2012

Mars Missions

Mariner 4

Destination: Mars flyby

Goals: Fly as close as 9,846 km to Mars, make measurements with various field and particle sensors and detectors, and take images.

Launch: November 28, 1964

Arrival: July 14, 1965

End of Mission: December 21, 1967

Phoenix Mars Lander

Destination: Mars

Goals: Study the history of water in the Martian arctic region. Search for evidence of past life in the ice-soil boundary. Analyze the water and soil for evidence of climate cycles and whether the environment could have supported microbial life. Take panoramic, stereoscopic images and close-up images of the soil and water ice from the Martian surface. Describe the geology of Mars, monitor the weather, and investigate properties of the atmosphere and clouds.

Launch: August 4, 2007

Arrival: May 25, 2008

End of Mission: Last communication November 2, 2008; officially ended May 24, 2010

Mars Pathfinder

Destination: Mars

Goals: Demonstrate key engineering technologies and concepts for use in future missions to Mars. Deliver science instruments to the surface of Mars to investigate the structure of the atmosphere, surface meteorology, surface geology, as well as the form, structure, and elemental composition of rocks and soil.

Launch: December 4, 1996

Arrival: July 4, 1997

End of Mission: September 27, 1997

Mars Global Surveyor

Acronym: MGS

Destination: Mars orbit

Goals: Circle Mars in a polar orbit (traveling over the north pole to the south pole and back to the north pole) once every 2 hours to collect global "snapshots" from 400 km above the Martian surface. Collect data to determine whether life ever arose on Mars, to investigate the character of the climate and geology, and to prepare for human exploration.

Launch: November 7, 1996

Arrival: September 11, 1997

End of Mission: Battery failure and spacecraft lost November 2, 2006; officially ended January 28, 2007

Mars Odyssey

Destination: Mars orbit

Goals: Map the amount and distribution of elements and minerals that make up the Martian surface. Record the radiation environment in low-Mars orbit to determine the radiation-related risk to any future human explorers.

Launch: April 7, 2001

Arrival: October 24, 2001

End of Mission: Still operating as of 2012

Mars Reconnaissance Orbiter

Acronym: MRO

Destination: Mars orbit

Goals: Search for evidence that water existed on the surface of Mars for a long time. Identify surface minerals. Study how dust and water are transported in the Martian atmosphere.

Launch: August 12, 2005

Arrival: March 10, 2006

End of Mission: Still operating as of 2012

Mariner 9

Destination: Mars orbit

Goals: Transmit images and measure UV and IR emissions.

Launch: May 30, 1971

Arrival: November 14, 1971

End of Mission: October 27, 1972

Mars Express

Destination: Mars

Goals: Search for subsurface water from orbit. Study the interaction between solar wind and the atmosphere of Mars. Find out what happened to the large amount of water that was once on Mars.

Launch: June 2, 2003

Arrival: December 26, 2003

End of Mission: Still operating as of 2012 (Planned to end December 31, 2014)

Viking 1 and 2

Destination: Mars

Goals: Obtain high-resolution images of the Martian surface. Characterize the structure and composition of the atmosphere and surface. Search for evidence of life.

Launch: *Viking 1*, August 20, 1975; *Viking 2*, September 9, 1975

Arrival: *Viking 1* orbiter, June 19, 1976; *Viking 1* lander, July 20, 1976; *Viking 2* orbiter, August 7, 1976; *Viking 2* lander, September 3, 1976

End of Mission: *Viking 1* orbiter, August 17, 1980; *Viking 1* lander, November 11, 1982; *Viking 2* orbiter, June 25, 1978; *Viking 2* lander, April 11, 1980

Mars Exploration Rover–Spirit

Acronym: MER

Destination: Mars

Goals: Study rocks and soils that might hold clues to past water activity on Mars.

Launch: June 10, 2003

Arrival: January 4, 2004

End of Mission: Last communication March 22, 2010; officially ended May 25, 2011

Mars Exploration Rover–Opportunity

Acronym: MER

Destination: Mars

Goals: Study rocks and soils that might hold clues to past water activity on Mars.

Launch: July 7, 2003

Arrival: January 25, 2004

End of Mission: Still operating as of 2012

Moon Missions

Lunar Prospector

Destination: Moon

Goals: Look for water ice buried inside the lunar crust and for other natural resources.

Launch: January 7, 1998

Arrival: January 11, 1998

End of Mission: Deliberate crash landing July 31, 1999

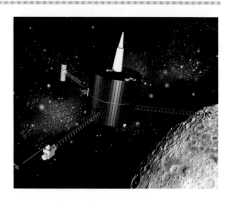

Surveyor

Destination: Moon

Goals: Test midcourse and terminal maneuvers and safely land on the Moon. Test the communications system during cruise, descent, and after landing. Look for locations that would be safe for Apollo landings.

Launch: *Surveyors 1* through *7*, from May 1966 to January 1968

Arrival: *Surveyors 1* through *7*, from June 1966 to January 1968

End of Mission: *Surveyor 7*, February 21, 1968

Ranger 7, 8, 9

Destination: Moon

Goals: Take high-quality images of the Moon and transmit them back to Earth in real time.

Launch: *Ranger 7*, July 28, 1964; *Ranger 8*, February 17, 1965; *Ranger 9*, March 21, 1965

Arrival and End of Mission: *Ranger 7*, July 31, 1964; *Ranger 8*, February 20, 1965; *Ranger 9*, March 24, 1965

Clementine

Destination: Moon orbit

Goals: Test lightweight imaging sensors and component technologies for the next generation of Department of Defense (DOD) spacecraft. Collect data on the Moon and asteroid Geographos.

Launch: January 25, 1994

Arrival: June 1994

End of Mission: June 1994

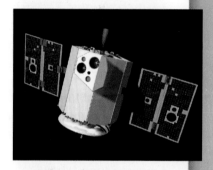

Apollo

Destination: Moon

Goals: Establish the technology to meet other national interests in space and to achieve preeminence in space for the United States. Explore the Moon and develop human capability to work in the lunar environment.

Launch: *Apollo 7*, October 11, 1968; *Apollo 17*, December 7, 1972

Arrival: *Apollo 17*, December 11, 1972

End of Mission: *Apollo 7*, October 22, 1968; *Apollo 17*, December 19, 1972

Lunar Crater Observation and Sensing Satellite

Acronym: LCROSS

Destination: Moon

Goals: Search for water in the form of ice in a crater that is always in shadow at the Moon's south pole. Deploy a device to crash into the crater, creating a plume of potentially icy debris to be measured, then crash itself into the crater, creating a second plume.

Launch: June 18, 2009

Arrival: June 23, 2009

End of Mission: October 9, 2009

Mercury and Venus Missions

Mercury Surface, Space Environment, Geochemistry, and Ranging

Acronym: MESSENGER

Destination: Mercury

Goals: Study the history, origin, and evolution of Mercury. Map nearly the entire planet in color, image the surface in high resolution, and measure the composition of the surface, atmosphere, and magnetic field.

Launch: August 3, 2004

Arrival: March 18, 2011

End of Mission: Still operating as of 2012

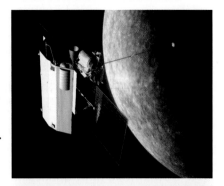

Mariner 2

Destination: Venus flyby

Goals: Collect data on Venus's atmosphere, magnetic field, charged-particle environment, and mass. Measure the solar wind and interplanetary medium.

Launch: August 27, 1962

Arrival: December 14, 1962

End of Mission: January 3, 1963

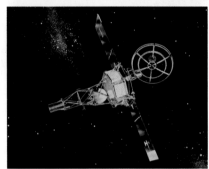

Mariner 5

Destination: Venus flyby

Goals: Measure interplanetary and Venusian magnetic fields, charged particles, plasmas, radio refractivity, and ultraviolet emissions of the atmosphere on Venus.

Launch: June 14, 1967

Arrival: October 19, 1967

End of Mission: November 1967

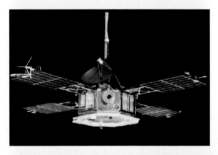

Mariner 10

Destination: Venus and Mercury flyby

Goals: Measure Mercury's environment, atmosphere, and surface characteristics, and make similar investigations of Venus.

Launch: November 3, 1973

Arrival: Venus flyby, February 5, 1974; Mercury flybys, March 29, 1974; September 21, 1974; March 16, 1975

End of Mission: March 24, 1975

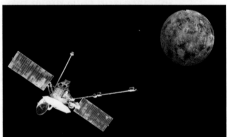

Venus Radar Mapper–Magellan

Destination: Venus orbit

Goals: Make detailed radar images of Venus, map the surface topography, and map the electrical characteristics. Measure Venus's gravitational field and show the planet's internal mass distribution.

Launch: May 4, 1989

Arrival: August 10, 1990

End of Mission: Descent into Venusian atmosphere October 11, 1994

Pioneer Venus

Destination: Venus orbit

Goals: Study the composition of the atmosphere and the characteristics of the upper atmosphere and ionosphere. Investigate solar wind near Venus. Map Venus's surface through a radar-imaging system.

Launch: Orbiter, May 20, 1978; Multiprobe, August 8, 1978

Arrival: Orbiter, December 4, 1978; Multiprobe, December 9, 1978

End of Mission: Descent into Venusian atmosphere: Orbiter, October 8, 1978; Multiprobes, November 15 and 19, 1978

Sun Missions

Hinode (Solar-B)

Destination: Sun orbit

Goals: Observe the Sun in high-resolution visible light. Observe solar magnetic fields. Investigate the heating mechanism of the solar corona with an X-ray telescope and extreme ultraviolet imaging spectrometer.

Launch: September 22, 2006

End of Mission: Still operating as of 2012

Reuven Ramaty High Energy Solar Spectroscope Imager

Acronym: RHESSI

Destination: Sun orbit

Goals: Investigate the physics of particle acceleration and energy release in solar flares. Observe the processes that take place in the magnetized plasmas of the solar atmosphere during a flare: impulsive energy release, particle acceleration, and particle and energy transport.

Launch: February 5, 2002

End of Mission: Still operating as of 2012

Genesis Solar Wind Sample Return

Destination: Solar wind

Goals: Collect samples of solar wind particles and return them to Earth for detailed analysis. Precisely measure the abundance of solar isotopes and elements, and provide a source of solar matter for future scientific analysis. Precisely measure ratios of oxygen, nitrogen, and noble gases in the different phases of solar activity.

Launch: August 8, 2001

Arrival: December 3, 2001

End of Mission: September 8, 2004

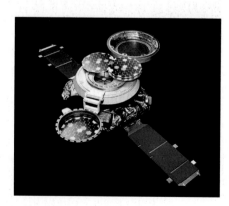

Investigation 9: Space Exploration

Solar and Heliospheric Observatory

Acronym: SOHO

Destination: Sun orbit

Goals: Answer three fundamental scientific questions about the Sun: What are the structure and dynamics of the solar interior? Why does the solar corona exist, and how is it heated to such an extremely high temperature? Where is solar wind produced, and how is it accelerated?

Launch: December 2, 1995

End of Mission: Still operating as of 2012

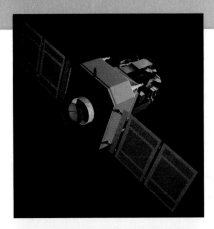

Transition Region and Coronal Explorer

Acronym: TRACE

Destination: Sun orbit

Goals: Explore the three-dimensional magnetic structures that emerge through the visible surface of the Sun, the photosphere. Define the dynamics of the upper solar atmosphere, the transition region, and the corona.

Launch: April 1, 1998

End of Mission: Last image taken on June 21, 2010

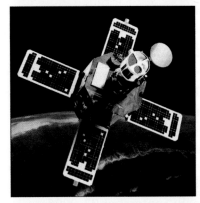

Solar Terrestrial Relations Observatory

Acronym: STEREO

Destination: Sun orbit

Goals: Trace the flow of energy and matter from the Sun to Earth. Collect data on the three-dimensional structure of coronal mass ejections (CMEs).

Launch: October 25, 2006

End of Mission: Still operating as of 2012

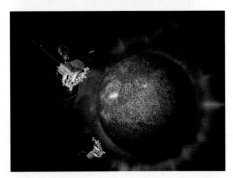

Solar Dynamics Observatory

Acronym: SDO

Destination: Sun orbit

Goals: Understand the causes of solar variability and its impacts on Earth. Study the Sun's influence on Earth and near-Earth space by observing the solar atmosphere at a small scale in many wavelengths at the same time. Determine how the Sun's magnetic field is generated, structured, and released in the form of energy. Predict solar variations that can affect life on Earth, such as solar flares that can affect cell phones and other technology.

Launch: February 11, 2010

End of Mission: Still operating as of 2012

Space Telescopes

Chandra X-Ray Observatory

Destination: Earth orbit

Goals: Observe X-rays from high-energy regions of the universe, such as the remnants of exploded stars, and active galaxies. Study the origin, evolution, and destiny of the universe.

Launch: July 23, 1999

End of Mission: Still operating as of 2012

Hubble Space Telescope

Acronym: HST

Destination: Earth orbit

Goals: Provide unprecedented deep and clear views of the universe, from the solar system to extremely remote young galaxies.

Launch: April 24, 1990

End of Mission: Still operating as of 2012

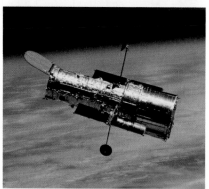

Kepler

Destination: Earth orbit

Goals: Survey our region of the Milky Way galaxy to detect and characterize Earth-sized and smaller planets in or near the habitable zone (within the distances from a star where liquid water can exist on a planet's surface).

Launch: March 6, 2009

End of Mission: Still operating as of 2012; expected duration 3.5 to 6 years

Spitzer Space Telescope

Destination: Earth orbit

Goals: Look into regions of space that are hidden from optical telescopes.

Launch: August 25, 2003

End of Mission: Still operating as of 2012

Investigation 9: Space Exploration

Exoplanet Transit Graphs

A: Kepler 10b

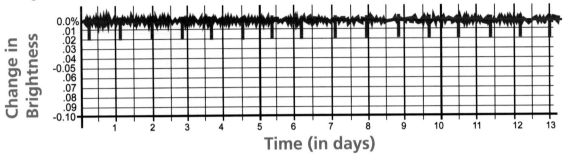

B: HAT-P-7b

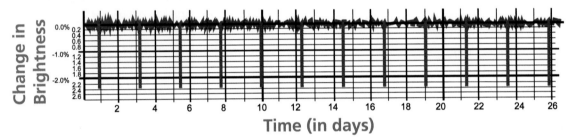

C: Kepler-11e, f, g

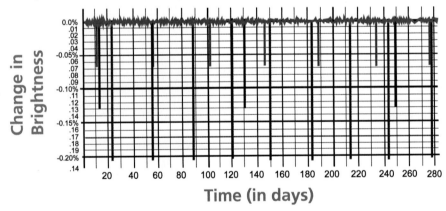

D: Mystery (Hint: there is more than one planet)

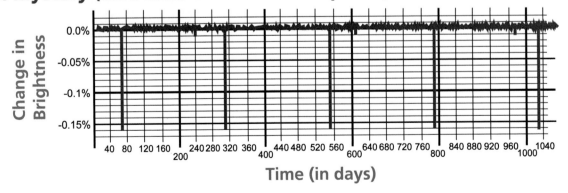

Based on Kepler Mission data collected in 2009–2011.

Science Safety Rules

1. Always follow the safety procedures outlined by your teacher. Follow directions, and ask questions if you're unsure of what to do.

2. Never put any material in your mouth. Do not taste any material or chemical unless your teacher specifically tells you to do so.

3. Do not smell any unknown material. If your teacher asks you to smell a material, wave a hand over it to bring the scent toward your nose.

4. Avoid touching your face, mouth, ears, eyes, or nose while working with chemicals, plants, or animals. Tell your teacher if you have any allergies.

5. Always wash your hands with soap and warm water immediately after using chemicals (including common chemicals, such as salt and dyes) and handling natural materials or organisms.

6. Do not mix unknown chemicals just to see what might happen.

7. Always wear safety goggles when working with liquids, chemicals, and sharp or pointed tools. Tell your teacher if you wear contact lenses.

8. Clean up spills immediately. Report all spills, accidents, and injuries to your teacher.

9. Treat animals with respect, caution, and consideration.

10. Never use the mirror of a microscope to reflect direct sunlight. The bright light can cause permanent eye damage.

Glossary

asteroid a small, rocky object that orbits the Sun

asteroid belt a region between Mars and Jupiter that consists of small chunks of matter that orbit the Sun

astronomical unit (AU) the average distance between Earth and the Sun, about 150 million kilometers

atmosphere a layer of gases that surround an object held in place by gravity

axis an imaginary axle that a planet spins on

big bang theory a theory that one explosion created the universe

black hole an extremely dense object that can form after a star goes supernova

circumference the distance around a circle

comet a chunk of ice, dust, and rock material a few kilometers in size

complex crater a crater that has central peaks and ejecta thrown out in long rays

cosmos the universe

crater a change in land created by an impact

crescent Moon the shape of the Moon just after and just before the new Moon phase

diameter the distance from one point in a circle to an opposite point in the circle

dwarf planet an object that orbits the Sun and is big enough to be round but doesn't clear away all objects near its orbit

ejecta material displaced from the land when a crater is formed

electromagnetic spectrum the range of electromagnetic radiation arranged in order of energy level

emit to give off

equinox a day of the year when the Sun's rays shine straight down on the equator

exoplanet a planet circling a star other than the Sun

first-quarter Moon the phase that occurs halfway between the new and full Moon

flooded crater a crater from a large impact that released magma from beneath the surface

full Moon the phase of the Moon that occurs when the Moon is opposite the Sun as seen from Earth

galaxy an enormous collection of tens of millions to hundreds of billions of stars, interstellar gas, and dust

gibbous a Moon shape that is larger than a first or third quarter Moon but smaller than a full Moon

gravity the force that causes two masses to attract each other

Kuiper Belt the region of the solar system beyond the orbit of Neptune; plutoids are located here

latitude the angular distance north or south from Earth's equator

light-year (ly) the distance light travels in 1 year. One ly is about equal to 9.5 trillion kilometers.

lunar eclipse when Earth is exactly between the Moon and the Sun, and the Moon passes through Earth's shadow

mare (plural **maria**) the dark surface of cooled magma in a flooded crater

meteor a streak of light in the sky from gravel- and pebble-size meteoroids; also known as a shooting star

meteorite a piece of a meteoroid that hits the ground

meteoroid a small or medium-size piece of rock or metal from space

Milky Way the name of the galaxy the Sun is a part of

Moon Earth's natural satellite

nebula (plural **nebulae**) a cloud of gas and dust in space between stars

new Moon the phase of the Moon that occurs when the Moon is in the direction of the Sun as seen from Earth

North Star the reference star pointed to by Earth's North Pole

orbit the path and length of time one object takes to travel around another object

orbit radius the average distance from one object to the object it is orbiting

orbital period how long it takes an object to orbit another object

parallel continuing in the same direction and always the same distance apart

phase each different shape of the Moon

planet an object that orbits a star and is massive enough for its own gravity to force it into a spherical shape

plutoid a type of dwarf planet that has an orbit beyond Neptune

revolution traveling around something

rotation spinning on an axis

satellite an object orbiting a larger object

season a period of the year identified by changes in hours of daylight and weather

simple crater a small, bowl-shaped crater that has a fairly uniform blanket of ejecta distributed around the rim

solar eclipse when the Moon passes exactly between Earth and the Sun

solar energy energy from the Sun

solar system a region of space occupied by a system of objects, including the Sun and all things orbiting it

solstice a day of the year when Earth's North Pole is leaning either toward the Sun or away from the Sun

spectroscope a tool used to study the spectrum of colors coming from a light source

star a large, hot ball of gas

star cluster a group of stars held together by their mutual gravitational attraction

Sun the star at the center of the solar system

supernova an explosion that ends a star's life

third-quarter Moon the phase that occurs halfway between the full and new Moon

transit when one object appears to move across another object as seen from the perspective of an observer

universe the sum total of all things that can be observed or detected

waning getting smaller

waxing getting bigger

Index

A
Alvarez, Walter, 37–40
Apollo missions, 31, 41, 43, 127
Asteroid, 32, 47, 63, 121, 134
Asteroid belt, 58, 63, 134
Astronomical unit (AU), 45, 134
Atmosphere, 31, 134
Axis, 10, 134

B
Barringer Crater (Meteor Crater), 31, 35, 42–43, 44, 96
Big bang theory, 53, 134
Binary star, 50
Black hole, 51, 134

C
Ceres (asteroid), 47
Chicxulub Crater, 39
Circumference, 9, 134
Columbus, Christopher, 3–6
Comet, 32, 42–44, 48, 67, 121–122, 134
Complex crater, 34, 134
Convection, Rotation, and planetary Transits (CoRoT), 74
Cosmos, 45, 46–48, 49–52, 56–57, 134
Crater, 31–44, 96–98, 134
Crescent Moon, 21, 22, 23, 134

D
Diameter, 35, 134
Dinosaurs, extinction of, 36–41
Dwarf planet, 47, 134

E
Earth, 4, 7–9, 10–13, 50, 56, 58, 61, 85, 88, 90, 99–107
Eclipses, 27–30
Ejecta, 33, 134
Electromagnetic spectrum, 68, 69, 134
Emit, 68, 134
Equinox, 11, 12, 134
Eratosthenes, 7–9
Eris, 66
Exoplanet, 71, 73, 74, 132, 134

F
First-quarter Moon, 25, 134
Flooded crater, 34, 134
Full Moon, 20, 25, 86, 134

G
Galaxy, 50, 52–53, 134
Galilei, Galileo, 64
Gas giant planet, 58, 63–65
Gibbous, 25, 134
Gravity, 32, 46, 56–57, 58, 72, 134

H
Hubble Space Telescope (HST), 71, 131

I
Incandescent lamp, 6
Incomplete circuit, 23, 75
Induced magnetism, 40, 75
Interact, 40, 75
International Morse Code, 67
Iron, 37, 40, 75

J
Jupiter, 32, 42, 47, 64, 70, 108–109, 114–115

K
Kepler Mission, 74, 75, 131
Kuiper Belt, 48, 58, 66, 134

L
Landforms, 99–107
Latitude, 5, 134
Light-year (ly), 45, 134
Lunar cycle, 21
Lunar eclipse, 27, 29–30, 134
Lunar month, 26
Lunar year, 20–21
Lunisolar cycle, 21

M
Mare, 34, 135
Mars, 32, 45, 46, 47, 54, 62–63, 112, 118, 124–126
Matter, 56–57, 69
Mercury, 46, 54, 60, 113, 119, 128–129
Meteor, 31, 48, 135
Meteorite, 37, 135
Meteoroid, 31, 33, 48, 135
Milky Way, 51, 52–53, 75, 76, 135
Moon (Luna), 14–18, 19–30, 31–44, 54–58, 61, 86–89, 90, 91, 92–95, 126–127, 135

N
Nebula, 49–51, 135
Neptune, 45, 46, 48, 48, 64, 111, 117
New Moon, 25, 135
North Star, 11, 135

O
Oort cloud, 32, 48
Orbit, 10, 135
Orbit radius, 75, 135
Orbital period, 75, 135

P
Parallel, 7, 135
Phase, 20, 135
Planet, 31, 46, 60–65, 71–76, 89, 108–119, 135
Pluto, 46, 47, 66, 67
Plutoid, 66, 135

R
Revolution, 10, 135
Rotation, 10, 135

S
Safety rules, 133
Satellite, 43, 46, 47, 90, 135
Saturn, 45, 46, 47, 48, 64, 110, 116–117
Season, 10–13, 85, 135
Shadow, 7–8, 9
Shocked quartz, 31, 40, 43
Shoemaker, Gene, 31, 42–44
Shoemaker-Levy 9 (comet), 32, 42, 67
Simple crater, 34, 135
Solar eclipse, 27, 28, 135
Solar energy, 12, 135
Solar system, 31, 46–57, 58–67, 68–70, 71–72, 89, 99–114, 122–124, 135
Solar year, 20–21
Solstice, 11, 12, 135
Space exploration, 68–70, 120–131
Spectroscope, 68, 135
Spectrum lines, 73
Star, 49–51, 135
Star blink, 74
Star cluster, 51, 135
Sun (Sol), 7–9, 10–13, 32, 46, 50, 54, 56, 58, 85, 89, 90, 129–130, 135
Supernova, 50–51, 135

T
Telescope, 71, 131
Terrestrial planet, 58, 60–63
Third-quarter Moon, 25, 26, 135
Tide, 57, 61
Transit, 74, 135

U
Universe, 76, 135
Uranus, 46, 65, 109, 116

V
Venus, 46, 54, 60–61, 113, 119

W
Waning, 26, 135
Water, hunt for, 68–70
Waxing, 25, 135
Wobble, 72–73